W9-COH-751

the GIS 20 essential skills

GINA CLEMMER

ESRI PRESS
REDLANDS, CALIFORNIA

Esri Press, 380 New York Street, Redlands, California 92373-8100

Copyright 2010 Esri

All rights reserved. First edition 2010

14 13 12 11 10   2 3 4 5 6 7 8 9 10 11

Printed in the United States of America

Library of Congress Cataloging-in-Publication Data

Clemmer, Gina, 1974-
  The GIS 20 : essential skills / Gina Clemmer. -- 1st ed.
      p. cm.
  ISBN 978-1-58948-256-2 (pbk. : alk. paper)
  1. Geographic information systems. 2. ArcGIS. I. Title.
  G70.212.C59 2010
  910.285--dc22   2010014268

The information contained in this document is the exclusive property of Esri unless otherwise noted. This work is protected under United States copyright law and the copyright laws of the given countries of origin and applicable international laws, treaties, and/or conventions. No part of this work may be reproduced or transmitted in any form or by any means, electronic or mechanical, including photocopying or recording, or by any information storage or retrieval system, except as expressly permitted in writing by Esri. All requests should be sent to Attention: Contracts and Legal Services Manager, Esri, 380 New York Street, Redlands, California 92373-8100, USA.

The information contained in this document is subject to change without notice.

U.S. Government Restricted/Limited Rights: Any software, documentation, and/or data delivered hereunder is subject to the terms of the License Agreement. In no event shall the U.S. Government acquire greater than restricted/limited rights. At a minimum, use, duplication, or disclosure by the U.S. Government is subject to restrictions as set forth in FAR §52.227-14 Alternates I, II, and III (JUN 1987); FAR §52.227-19 (JUN 1987) and/or FAR §12.211/12.212 (Commercial Technical Data/Computer Software); and DFARS §252.227-7015 (NOV 1995) (Technical Data) and/or DFARS §227.7202 (Computer Software), as applicable. Contractor/Manufacturer is Esri, 380 New York Street, Redlands, California 92373-8100, USA.

Esri, www.esri.com, the Esri Press logo, @esri.com, ArcGIS, ArcView, ArcMap, ArcCatalog, ArcToolbox, ArcSDE, SDE, and Spatial Database Engine are trademarks, registered trademarks, or service marks of Esri in the United States, the European Community, or certain other jurisdictions. Other companies and products mentioned herein are trademarks or registered trademarks of their respective trademark owners.

Ask for Esri Press titles at your local bookstore or order by calling 800-447-9778, or shop online at www.esri.com/esripress. Outside the United States, contact your local Esri distributor or shop online at www.eurospanbookstore.com/esri.

Esri Press titles are distributed to the trade by the following:

In North America:

Ingram Publisher Services
Toll-free telephone: 800-648-3104
Toll-free fax: 800-838-1149
E-mail: customerservice@ingrampublisherservices.com

In the United Kingdom, Europe, Middle East and Africa, Asia, and Australia:

Eurospan Group   Telephone: 44(0) 1767 604972
3 Henrietta Street   Fax: 44(0) 1767 601640
London WC2E 8LU   E-mail: eurospan@turpin-distribution.com
United Kingdom

CONTENTS

ACKNOWLEDGMENTS

This book would not have been possible without the fabulous ESRI team that shaped and polished these chapters. Judy Hawkins, acquisitions editor, skillfully and gently guided me across uncharted terrain, helped me find my footing, and cheered me on. Claudia Naber, my editor, helped me express myself more clearly and ironed out many wrinkles in the text. Riley Peake, cartographer, applied his mad mapping skills to ensure fluidity within the exercises by checking and rechecking the steps. He was also a great sounding board, and so polite. Brian Harris made permissions as painless as possible. Monica McGregor, handling graphics and layout, made some fairly dry text look great.

To my tenacious team at New Urban Research Inc., thank you for doing many things I should have been doing while instead I was writing this book. Your input, support, and laughter are invaluable to me. Thanks for keeping our little boat afloat.

Thank you to Joan Lufrano for giving me the idea in the first place.

Alan Peters introduced me to GIS at the University of Iowa, where he was a professor in the Urban and Regional Planning program. He saw the value of GIS when its virtues in the planning community were still in debate. By making GIS accessible and teaching us how to use the darn thing, he changed the course of my life. I am so lucky that I was his student.

Thank you to my husband, Richard Lufrano, for spending many hours editing this book. Your edits not only clarified things, but reinforced the foundation of this book. More importantly, your support and confidence in me, cheering me on at every lap, helped me make it to the finish line. I couldn't ask for a better partner and friend. To the rest of our pack: Merlin, Max, Joye, Rick, Gerri, Nanny, Kim, Joan and Ned, Mike and Lyz, thanks for all the love.

PREFACE

Over the past decade, it has been my passion to provide an applied approach to teaching GIS to busy professionals. My pedagogy is different from others. I believe the fundamentals of GIS (and ArcGIS specifically) can be taught

- quickly. I do not believe that GIS is the deeply complicated discipline many make it out to be.

- using a project-based paradigm by completing a concrete task as the goal, versus a layered or building block approach. In other words, I don't think you need to know the inner workings of the GIS to create common types of maps.

- by learning what is most frequently used by most people (and therefore most important to understand), and skipping the rest. I do not believe that you must know every aspect of ArcGIS in order to successfully complete common GIS tasks.

- using everyday language versus technical jargon.

This book is an extension of my passion to help professionals new to GIS quickly learn the essential fundamentals of ArcGIS. The purpose of the book is to provide a focused approach to learning GIS by offering clear, easy-to-follow exercises for twenty of the most commonly used GIS skills in the industry today.

I have spent most of the past decade training thousands of new GIS users. I created a workshop for busy professionals called *Mapping Your Community: An Introduction to GIS and Community Analysis*. It has been taken by more than 20,000 working professionals. At the end of each class, students are asked to provide feedback about the course and its content. These student evaluations helped shape my class framework and contributed mightily to the content of this book.

In addition to the feedback my students offer, in October 2008 I conducted a survey of 500 GIS professionals to determine and quantify the top twenty GIS skills currently used by working professionals. This book is a direct result of that research. The chapters that follow reflect the survey's results, as well as my empirical knowledge of the GIS industry.

Gina Clemmer
New Urban Research Inc.
April 2010

INTRODUCTION

You can read this book in sequence or skip to the chapters that target the skills you want. How to get the data for all exercises is discussed in the beginning of each chapter.

To get started using this book, you need software, shapefiles, and data—all introduced below. Then we cover helpful resources, tools, and tips.

ArcGIS Desktop software

You must have an installed copy of ArcGIS Desktop with an ArcView license to complete these exercises. No special extensions are necessary. It is intended that the reader be able to accomplish these exercises with minimum software requirements. ESRI is the industry leader in desktop geographic information system (GIS) technology and, while there are many other companies that make GIS applications, given the popularity and functionality of ArcGIS, we chose this software to illustrate the GIS 20.

This book was written using the latest version of the software, ArcGIS 10. You can use older versions of the software (8.x and higher) to accomplish most things in this book. If something is absolutely not doable in older versions, it will be noted in the exercise.

ArcGIS installation

- ArcGIS only works on a PC, not a Mac.

- Memory requirements are 1 GB minimum, 2 GB recommended or higher, and the recommended processor speed is 1.6 GHz.

- A free 60-day trial copy of the software can be ordered from http://www.esri.com/software/arcgis/arcview/eval/evaluate.html.

Licensing

- Licenses come in two types: single-use and concurrent. You can purchase single-use licenses intended for one user, or concurrent licenses intended for a network that supports up to three occasional users. This does not mean three people can access the software at once; only one person can access it from the network at a time.

- For a single-user license the cost of the software is $1,500 (as of April 2010). You may purchase the software here: http://www.esri.com/software/arcgis/arcview/pricing.html.

Discounts are offered for

- ordering two or more copies (the more you order the bigger the discount)
- those in education (faculty, students, or staff)
- federal employees
- municipalities and other large organizations that have a Master Purchasing Agreement (MPA) or Enterprise Licensing Agreement (ELA)

Nonprofit grants are also available but mostly for conservation programs. A better route might be to look at another option. ESRI has an agreement with TechSoup.org to provide discounted ArcGIS licenses to nonprofits ($195). Only one license per organization per year is allowed.

Shapefiles

The exercises in this book use shapefiles. Shapefiles are like basemaps and are widely used throughout the GIS industry.

One of the most difficult decisions in writing this book was whether to include getting shapefiles as part of the exercise, providing shapefiles on a CD, or letting you work with your own shapefiles. We decided to do all three.

In each chapter you have the choice of doing the exercise as written, which, in most instances, will walk you through downloading required files. This is helpful if you want to learn how to access commonly used public files from the U.S. Census Bureau. This is what people who use GIS do.

If you simply want to learn the GIS software part and you already have a good idea about how to access files, then you can use the CD that includes all files needed to do the exercises.

You may even have your own shapefiles, and we give you information about how you can substitute them to do the exercises. Each chapter provides descriptions for the types of files needed to complete each exercise. You may find this alternative useful and immediately applicable to your work.

Hunting down and integrating your own shapefiles not only will make the exercises more relevant to your immediate work, it will also encourage you to begin thinking about where to obtain current shapefiles for your other projects, which is central to this book's theme of "keeping it real." Finding and working with shapefiles from disparate sources is as real as it gets, and something that those in the GIS world do daily.

CURRENCY OF SHAPEFILES

One key consideration with shapefiles is knowing how current they are. Determine the year and month to which your shapefiles pertain. The more current the better.

Shapefile resources

The U.S. Census Bureau is our nation's custodian of the legal geographic definitions of United States' borders. It is the largest distributor of shapefiles. The bureau used to maintain geographic boundaries in TIGER/Line files format but recently it has switched to shapefile format. Shapefiles are updated at least once a year and are as current as the previous year.

The U.S. Census Bureau is the best resource for free, current shapefiles. To download shapefiles from the census, go to http://www2.census.gov/cgi-bin/shapefiles2009/national-files.

Other places to get shapefiles include the following:

- ESRI Data & Maps DVD that comes with the ArcGIS software. You can read more about what is included here: http://www.esri.com/data/data-maps/index.html. One of the most helpful things is that this DVD includes many non-U.S.-based shapefiles.

- Shapefiles and current ESRI data can be instantly downloaded at the fee-based site Primary Data Source: www.primarydatasource.com.

- Many organizations are building GIS departments and have shapefiles available internally.

- If you are still unable to find what you need, search the Internet for your location using the word "shapefile" or just the extension ".shp."

Data

Exercises in this book require you to use your own data. Where relevant, exercises use Microsoft Excel to manipulate and organize data. Other data programs can be used including SPSS, SAS, and Microsoft Access.

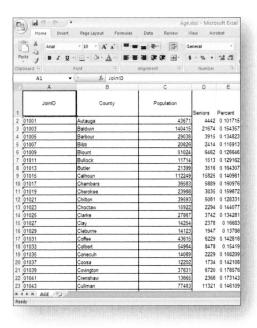

In general, data should be saved as one of the following formats to work well in ArcGIS:

- Microsoft Excel (.xls)
  - Excel files only work in 9.3 and higher. For older versions of the software, you must save Excel spreadsheets as a database file (.dbf) file type.
  - Older versions of ArcGIS do not read Excel 2007 files (.xlsx). In Excel 2007, you must save it as an older .xls version.
- Database File (.dbf)
  - Excel 2007 doesn't allow you to save as a .dbf; depending on what you are doing in ArcGIS, you may be able to save as a .csv.
- Comma Separated Values (.csv)
- Microsoft Access (.mdb or .accdb)

One of the best sources for free demographic information is via the census. See chapter 4, "Preparing data for ArcGIS" for information on downloading census data.

Other important elements to your success with the GIS 20 include familiarity with the ArcGIS help text, basic and essential tools, everyday language instead of technical jargon, and a few incredibly useful tips.

ArcGIS help menu

Since this book does not cover every single aspect of ArcGIS, learning to use the help menu will be, well, helpful. Two help menus are built into the ArcGIS interface. One comes with the software and is a static body of work, meaning that it is not updated as time goes on. To help you to easily find the most relevant information related to specific topics, watch for the ⦿ symbol for key help menu search terms included throughout the book. ❶

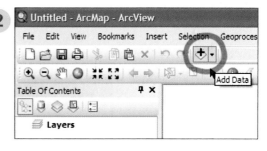

The second help option is the online ESRI Resource Centers. The great thing about the Resource Centers site is that it is frequently updated and loaded with new information. Be sure not to let all the information overwhelm you.

Essential tools

Because ArcGIS provides hundreds of tools, it's essential to identify the most important ones. The tools that you will use in nearly every mapping session are featured below. It's a good idea to become very familiar with these tools and what they can do.

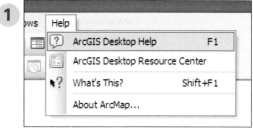

Adding layers

The fastest way to add geography files or data tables to your map is to select the Add Data icon on the Tools toolbar. ❷

Magnifying/demagnifying

There are four methods for zooming in and out of your map.

Panning

The Pan button allows you to reposition your map when you are in data view.

Going to the full extent

This tool will resize your map so it all fits onto your screen. It is a great way to center your map.

Default pointer

The default pointer doesn't do anything. That's the beauty of it. (One good thing to know is that to "deactivate" any of the other tools, you click the default pointer. This gets rid of the first tool and activates the default pointer.)

Identify

You can use the Identify tool to click specific geographies and look at the underlying data.

ArcCatalog

ArcCatalog is like the library system of ArcGIS. You can browse, copy, delete, and organize files here. 3

Search window

The Search window allows you to easily find all available tools.

Language

This book seeks to explain ideas and steps in nontechnical, everyday language using terms that the average person would be able to understand, not in GIS jargon.

What is the difference between ArcGIS, ArcMap, and ArcView?

ArcGIS, ArcMap, and ArcView are often used interchangeably. This book uses the terminology ArcGIS to refer to the entire mapping software suite used to complete exercises. ArcMap is the browser for the software, it's not the whole software. For example, Internet Explorer and Firefox are browsers to the Internet, they are not the Internet itself. This is what ArcMap is to ArcGIS. ArcView is the licensing level of the software. This software used to be called ArcView, but in 1999 ESRI made many enhancements and changed the name to ArcGIS.

Exercises in this book seek to use the most basic version of the software, which is ArcGIS Desktop with an ArcView-level license. Many other types of licenses and extensions are available for specialized tasks. These are not used in this book.

Incredibly useful tips

Right-click > Properties

If there is one ArcGIS tip that can make your life easier, it is to right-click on the layer name in the table of contents and select Properties. This will give you access to all available options for that layer. This right-click > Properties trick also works when creating a layout. If you select any element (the map, a title, the legend, the scale bar) and right-click that element and navigate to Properties, you'll gain access to all available options for that element.

Connect to folder

When beginning to work on a project, it is helpful to "connect to folder," which creates a permanent shortcut to any specified folder. You create the shortcut through ArcCatalog. By connecting to folder, you don't have to navigate through a maze of links every time you want to open a shapefile or project.

How to connect to folder from ArcCatalog

1. Open ArcGIS.
2. Click ArcCatalog.
3. You can Connect to Folder in two ways. The first way is to click the Connect to Folder icon. **4** Then navigate to folder to establish a shortcut. A common option is to "connect to folder" for the desktop. You don't actually have to choose a folder; you can just navigate to the desktop and stop there.
4. The second way is to right-click Folder Connections, then select Connect to Folder and navigate to the folder.

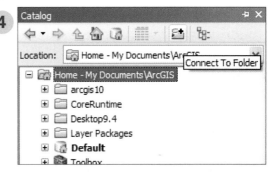

Accessing connected folders (or how do I find my files?)

1. Open ArcGIS.
2. Click the Add Data icon. ✛ ▾
3. Use the drop-down menu to select Folder Connections link. All connected pathways should be evident here. **5**

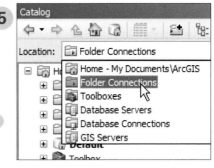

CHAPTER 1

Creating a reference map

Reference maps are basic, traditional maps. Their
purpose is to illustrate geographic boundaries for cities,
counties, and other areas. Reference maps have no
underlying data, such as population numbers, associated
with its layers. They are *just* geography as illustrated
in the map of Alabama that follows. Typically, reference
maps show multiple geographic layers.

This reference map ① shows Alabama county boundaries and Alabama city names with no outline around the cities. Labeling cities and not shading them gives a general idea of where the city is located, without putting so much stuff on the map that it's hard to read. The county and city layers are separate.

Reference mapping is an essential GIS skill, and a good place to begin your GIS training.

Exercise goal

Create a map that clearly displays multiple geographic layers.

A key point of this exercise is to teach you to effectively color shade and label *multiple* layers in a map. In this example, we're using counties and cities. For a reference map, it is difficult to illustrate more than four layers on one map without the map becoming too cluttered.

Exercise file locations

Determining where to get shapefiles can be confusing. Three options are shown below to help you decide where and how to get files for this exercise.

Chapter directions: Follow the exercise as it appears in this book
The shapefiles for this exercise were obtained from the U.S. Census Bureau Web site. Instructions on how to download these files are included as a part of this exercise in steps 1-3.

The state used in this exercise is Alabama and the layers are

- County and Equivalent (Current)
- Place (Current)

Note: These files will be used in several subsequent exercises.

CD: Use the CD included with this book
All files needed for this exercise are included on the book's CD. Files are organized by chapter.

Personal files: Use files you've gathered from other sources
You may want to create a map of some geography other than Alabama.

Perhaps you would like to create a reference map for a current project you are working on. For example, you might want to create a map of a specific neighborhood, ZIP Code, school district, or a customized boundary. To do this you will need to have access to those shapefiles ahead of time.

Because a key point of this exercise is learning to layer multiple layers, you will need at least two shapefiles, but not more than four. If your files are downloaded and unzipped, skip to step 5.

Downloading files

One of the best resources for current and free shapefiles is the U.S. Census Bureau. This chapter requires at least two shapefiles. We have chosen counties and cities because these two geographies are frequently used and helpful in showing you how to layer and label shapefiles, and it's easy to download these two files for any state.

1 Select files from the U.S. Census Bureau Web site

1. Navigate to www.census.gov.
2. On the main Census site, to the right of the Geography link, select the TIGER link. A link on that page is the gateway to the most current Census shapefiles. 2
3. Select the 2009 TIGER/Line Shapefiles Main Page link.
4. On the next page, select the Download Shapefiles link on the left, in the orange TIGER Navigation panel.

2 Select geography

Instead of a layer for the entire nation (as you would get if you selected a link on the left), you only want shapefiles for Alabama. Select Alabama from the drop-down menu on the right and click Submit. (Do not click any check boxes on the left.) 3

After selecting Alabama as the state, you will see a list of all available shapefiles for Alabama. Here is where things get a little confusing,

Alabama Counties and Cities, 2009

Legend

Counties

Map Source: New Urban Research, Inc. 2009.
Data Source: 2008 Census TigerLine Shapefiles, December 2008 Release.
Retrieved August 19, 2009, from http://www.census.gov: http://www2.census.gov/cgi-bin/shapefiles/state-files?state=01.

Jobs@Census
Catalog
Publications
Are You in a Survey?
About the Bureau
Regional Offices
Doing Business with Us

Business & Industry — Economic Census · Get Help with Your
Economic Indicators · NAICS · Survey
Government · E-Stats · Foreign Trade
Local Employment Dynamics · More

Geography — Maps · TIGER · Gazetteer · More

Newsroom — Releases · Facts For Features · Minor

State- and County-based Shapefiles

To download state-based shapefiles, select a state or equivalent.

To download county-based shapefiles, select a state, and then a county.

Alabama ▼   submit

mostly due to a technical problem on the census site. Theoretically, you can check the check box next to the two (or more) layers you want (County and Equivalent (Current) and Place (Current)) and click the Download Selected Files button at the bottom of the screen.

The files will download as one zipped file that contains multiple zipped files. Herein lies the problem. If you are using the unzipping program that comes standard with Windows XP, when you try to unzip the folder containing multiple zipped files, you will get an error message that says *Windows has blocked access to files to help protect your computer.* Use the steps below to fix this problem.

3 Download files and unzip

1. Click the County and Equivalent (Current) link. ④

2. When asked whether you want to Open, Save, or Cancel the file, click Save to save the zipped file on your C drive.

3. Repeat these steps for the Place (Current) file.

4. Unzip files. You can use any unzipping program you wish. Follow the instructions for that program to unzip the

④

Alabama Shapefiles

Select All	Clear Selection

- ☐ American Indian/Alaska Native/Native Hawaiian Area (Current)
- ☐ American Indian/Alaska Native/Native Hawaiian Area (Census 2000)
- ☐ Block (Current)
- ☐ Block (Census 2000)
- ☐ Block Group (Census 2000)
- ☐ Census Tract (Census 2000)
- ☐ Congressional District (111th)
- ☐ Congressional District (108th)
- ☐ County and Equivalent (Current)
- ☐ County and Equivalent (Census 2000)
- ☐ County and Equivalent (Economic Census)

files. Each file should unzip in its own folder. If you are unsure whether your computer has an unzipping program, try double-clicking on the zipped file to unzip it. By downloading files individually, you get around issues about multiple zipped files.

Creating a reference map

4 Add shapefiles

1. Open ArcGIS by double-clicking the ArcMap icon on your desktop. If you do not see an icon, access the program from the Start menu by going to All Programs > ArcGIS > ArcMap 10.

2. Add shapefiles to the ArcGIS window by clicking the Add Data icon ✛ ▾.

3. Select the Connect to Folder icon 📁 and navigate to the folder where your files are stored, then select the shapefiles you downloaded (**tl_2009_01_county** and

STACKING LAYERS

When you have multiple layers in your map, the layers are stacked one on top of the other, and the order they're stacked in makes a difference. Whichever layer appears first in the table of contents displays at the top of the map. When this layer has a solid fill, you will not be able to see the underlying layers. We will fix this in step 7.

tl_2009_01_place). You can add more than one file at a time by holding down the Ctrl key on your keyboard and selecting multiple files. The path and folder you connect to will be stored under the Folder Connections section in the Add Data drop-down menu for easy selection later.

5 Turn layers off and on and reorganize them (optional)

1. You can display layers on your map by checking the check box next to each layer to turn it on, or by unchecking it to turn that layer off.

2. You can move layers up or down within the map and table of contents by highlighting the layer in the table of contents (clicking it once) and dragging it to the desired position.

3. Drag **tl_2009_01_county to** the first position in the table of contents (if it is not already in first position — just to practice moving layers).

6 Change layer names

1. Left-click *once* on the layer name to activate the text. Type over the existing layer name. Ultimately these names will appear on your created legend.

2. Rename layers so they will make sense to anyone who looks at the map, using common names like Counties and Cities.

7 Change layer colors (called changing the symbology in ArcGIS)

You'll need to change the layers' symbology (color / fill / outline) so that you can easily see each layer in the map.

For the county layer, follow these steps:

1. In the table of contents, right-click Counties.

2. Select Properties.

3. Select the Symbology tab.

4. Click the big color patch, under Symbol.

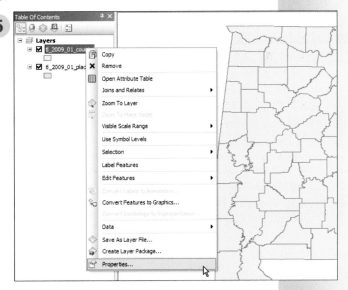

5. Click a colored square to change the color of this layer. In this example, the color should be Hollow. **6**

6. The Outline Color, slightly below the Fill Color option on the right, should be a medium gray. Gray is usually a better choice than black for the outline because it puts less ink on the page, which will make your maps easier to read. ⑦

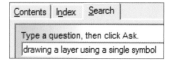

7. Click OK twice.

Hollow makes the layer transparent so you can see the layer underneath it. Hollow and No Color on the color palette are one and the same.

For the cities layer, follow these steps:

1. In the table of contents, right-click Cities.

2. Select Properties.

3. Select the Symbology tab.

4. Click the big color patch, under Symbol.

5. Click Hollow as the fill color *and* No Color as the outline color. This will remove all color for the Cities layer. That's just what we want to do, because in the next few steps we'll apply labels to the cities to show where each is located. This is a neat cartography trick, because it allows us to show approximately where the cities are but without using an outline.

6. Click OK twice.

Before rushing to the next step, practice changing and filling outline colors of any layer. You may want to practice with other colors to see what you think. In general, a good guideline is to make Hollow the fill color of most layers, but to assign each layer its own outline color.

6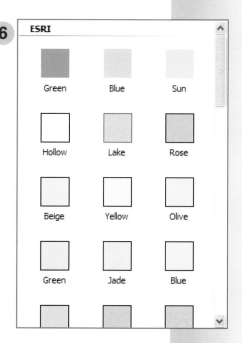

THE TWO MOST POPULAR TABS

Symbology and Labels are the two most frequently used property tabs. It is well worth your time to get to know how each works.

8 Turn on labels

Labeling is essential to creating a quality reference map.

1. In the table of contents, right-click the layer that you would like to label (Cities).

2. Click Properties and select the Label tab.

3. Check the Label features in this layer check box (a tiny little check box in the upper left corner) and select the appropriate field to use for labeling (Name). **7** ⑦

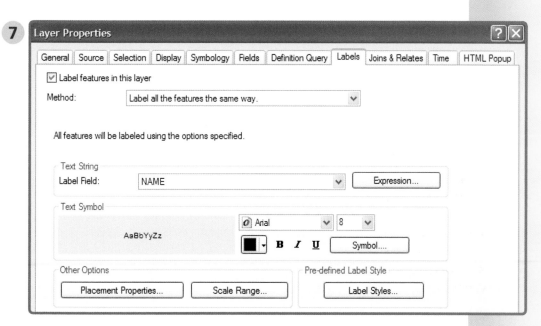

4. Click OK twice and have a look at the labels.

9 Fix labels

If your labels look messy, there are three things you can do to improve them:
1. Remove duplicate labels.
2. Assign a buffer to labels, which places only some labels.
3. Create a halo (a white outline) around labels.

1. In the table of contents, right-click Cities.
2. Click Properties and select the Labels tab.
3. Select Placement Properties in the lower left corner.
4. Click the Placement tab at the top.
5. Select the Remove Duplicate Labels option.
6. Switch to the Conflict Detection tab at the top.
7. Down toward the bottom of the menu there is an option to assign a buffer number. Type **2** in the provided field and click OK twice. (This will place *some* of your labels on the map, and will make your map much easier to read.)

8. To place a halo around labels, right-click Cities, select Properties, and then select the Label tab.

9. Click the Symbol button to make changes to label fonts.

10. Click B to make the labels bold.

11. To apply the halo, click the Edit Symbol button and select the Mask tab.

12. Select the halo option, then click OK three times.

10 Create a layout and insert map elements

As a general rule, it's a good idea to create a layout before you print. The layout includes all map elements such as a legend, north arrow, scale bar, and title. Most of the time you will work in Data View, but before printing you will need to switch to Layout View.

1. On the View menu, select Layout View. The window should resemble a piece of paper with your map on it.

    NOTE: A larger discussion of layout considerations can be found in the next chapter.

2. Click the Insert drop-down menu and select map elements such as a title, legend, north arrow, scale bar, and text (to add a source or footnote). To edit any of these elements, remember the incredibly useful shortcut of right-clicking the element and clicking Properties.

11 Export the map image

To export your map layout as a JPEG (.jpg) or Adobe Portable Document File (.pdf) do the following:

1. On the File menu, select Export Map.

2. Specify where you want to save your file on your C drive.

3. Specify the file format in the Save as type field.

    NOTE: For more information on exporting more complex PDFs, see chapter 19.

12 Save the project

ArcGIS saves projects like these as a map document with an MXD extension (.mxd). Saving an MXD file is like saving a whole project at once. Think of an MXD file as your entire workspace.

1. On the File menu, click Save As.

2. Navigate to where you would like to save on your C drive.

3. Type a name for this project.

CHAPTER 2

Creating well-designed layouts

ArcGIS allows you to view maps in Layout View or Data View. You can access both views from the View menu. When creating a map, it makes sense to use Data View because it provides the best way to work with all of the map's layers and features. However, prior to printing your map you must create a layout and insert all the elements that make a map look like a map (legend, title, scale bar, and north arrow).

Map elements such as a title, legend, north arrow, and scale bar should be added to the map prior to printing. This is accomplished in Layout View. You will spend about 10 percent of your time in Layout View (unless you have a really complicated layout). Think of Data View like a working document, and Layout View like a print preview. **1**

Exercise goal

Create a well-designed map layout.

Exercise file locations

Chapter directions: Follow the exercise as it appears in this book

The shapefile used in this exercise is for Alabama counties. It is the same file used in chapter 1.

County and Equivalent (Current)

If you did not download this file in chapter 1, you can follow the instructions from chapter 1.

CD: Use the CD included with this book

All files needed for this exercise are included on the book's CD. Files are organized by chapter.

Personal files: Use files you've gathered from other sources.

To complete this exercise you can use any shapefile.

1 Add shapefiles and switch to Layout View

1. Open ArcGIS by double-clicking on the ArcMap icon on your desktop. If you do not see an icon, access the program from the Start menu.

2. Add a shapefile to the ArcGIS window by clicking the Add Data icon. ✚ ▾

3. Either open Folder Connections or select the Connect to Folder icon and navigate to **tl_2009_01_county**, which was downloaded in chapter 1.

4. On the View menu, select Layout View. The window should resemble a piece of paper with your map on it.

2 Change page orientation and resize the map (optional)

Depending on whether the geography you are mapping is horizontal (Nebraska) or vertical (California), change the map's orientation to accommodate the geography, either portrait or landscape. To change the orientation from the default, portrait, to landscape, follow these steps:

1. Right-click anywhere outside of the map area.

2. Select Page and Print Setup.

THE PRINT AREA

In Layout View, the print area is defined by the default square. This square represents an 8 1/2-by-11-inch piece of paper. Stay within the boundaries of this square. Anything outside the boundaries will not print.

3. Select the Landscape option.

4. Select the Scale Map Elements check box (lower right) to center your map in the Page Layout.

5. Click OK.

6. Once the orientation changes to landscape, resize the image to fit the space. You can easily do this by clicking the map once to select it, then clicking and dragging the dots in each of the four corners. Don't worry about skewing your map when you drag it — ArcGIS keeps the correct proportions of your geography. Click one of the

LANDSCAPE OR PORTRAIT

Most of the time, a landscape page format will allow for a larger map image.

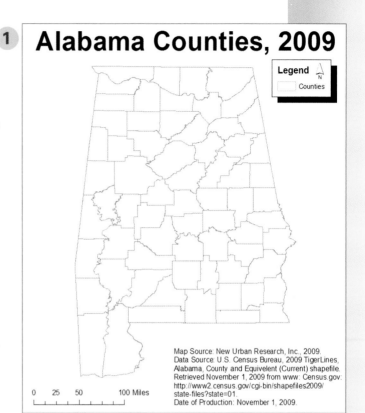

little dots on the corner until you get a double-headed arrow. Drag the image to the desired position and repeat with each corner dot until the map is as large as you would like it to be.

3 Understand the relationship between Data and Layout Views

What you see, in terms of the size of your map in Layout View, is what you have set up in Data View. To better understand this relationship, let's change the size of your map in Data View and see how that affects Layout View.

1. Go back to Data View (on the View menu) and use the Fixed Zoom Out tool ⟲ to make the map much smaller. Click it 6 or 7 times until your map is very small.

2. Switch back to Layout View and notice that what you see here is a visualization of Data View.

3. Switch back to Data View and use the Fixed Zoom In tool ⟲ to resize your map.

4. Return to Layout View.

RESIZING THE MAP

Your map may look too large in Data View, but keep switching between Data and Layout View until the map is sized to your specifications.

4 Change map color

1. Switch back to Data View.

2. In the table of contents, right-click the shapefile name, then select Properties and the Symbology tab.

3. Click the big color patch, under Symbol.

4. Click a colored square to change the color of this layer. In this example, the color should be Hollow.

5. The Outline Color (slightly below where the Fill Color option is) should be a medium gray. Gray is usually a better choice than black for the outline because it puts less ink on your page, which will ultimately make your maps easier to read.

6. Click OK twice.

7. Switch back to Layout View.

5 Insert title

A title must include three things: variable name, year of the data or geography, and geography type (even though the geography type like State may be obvious).

1. Switch to Layout View if you are not already there (the following steps can only be done in Layout View).

2. From the Insert menu, select Title.

3. Type **Alabama Counties, 2009** then click OK. **2**

4. Move the title box above the map by holding down the mouse's left-click button and dragging the box to the top of the page.

5. To change the font, right-click the title box.

6. Select Properties.

7. Select Change Symbol. Here you can change the font type, size, and style. Title fonts should be 16-point or higher. Change the font size to 36 and make it bold. Notice you can also select one of the font templates on the left.

8. Click OK twice.

   One quick way to get the right title size is to use the Fit Width to Margins option. Right-click the title. Select Distribute, then Fit Width to Margins.

INSERT MENU: A ONE-STOP SHOP

You can add all map elements from one place: the Insert menu.

TITLE PHRASING

When displaying a map with data, a great way to phrase a title is by using the phrase "Distribution of ..." such as Distribution of Poverty Rates by State, 2006. This phrasing works well with thematic, categorical, and geocoded maps.

FONTS

When creating a map, fonts matter. Of the two types of fonts, serif and sans serif, sans serif is easier to read. Common examples of sans serif fonts include Arial and Helvetica, both of which are smart choices for map fonts. Choose one font for all the text in your map.

6 Insert legend

1. From the Insert menu, select Legend.
2. Click Next then Next again to proceed through the menu. The default options work fairly well, except for the background color. ③
3. On the Legend Frame wizard, select a 0.5 border for the first option. For the Background color, select White. The default for a legend is Hollow, which makes the Legend see-through. A better choice is a solid fill of White.
4. For a drop shadow, select Black (optional).
5. Click Next, then Next again, and Finish.
6. Move the legend to the upper right corner of the map. This is usually a good area for the legend.

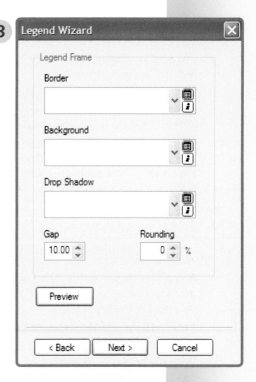

GOOD TO KNOW ABOUT LEGENDS

In this map we only have one layer. When creating a map with multiple layers, the legend should be organized in order of scale (from largest to smallest, for example, such as state, county, and city). Another good idea is to put the land masses (physical land like state, county, city) at the top of the legend and other context layers, like streets and water, at the bottom of the legend.

7 Insert scale bar

Traditional cartography instructs that scale bars must accompany every map. However, scale bars are frequently left off thematic maps (chapter 6) because they serve no real purpose. Scale bars are required on reference maps, since distance is an important element.

1. From the Insert menu, select Scale Bar.
2. Select the first scale bar template by clicking it once, and then selecting Properties.
3. Change Division Units from Decimal Degree to Miles. ④
4. Click OK twice.
5. The scale bar is always dropped right in the middle of the map. Move the scale bar to the lower right-hand corner.

CHANGING LAYER NAMES

If you want to change the name of the layer, click once on the layer name in the table of contents and edit. That will change the layer name in the legend.

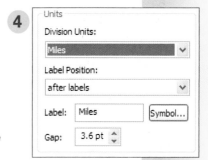

6. The easiest way to change the mileage of the scale bar is to drag the right side dot of the scale bar box to widen the scale bar. If you change the size of your map, the scale bar automatically adjusts. It's not a fixed element; it will change if you change the size of your map. Widen the scale bar to show 100 miles.

8 Insert source

As a cartographer, you can provide sources for data, files, and maps. The idea behind citing data sources is to help the reader verify the primary source of data. A good mapmaker gives the reader enough information to do so. Providing data sources is a must, but you may also consider providing sources for shapefiles as well as the map itself.

There are many ways to cite sources. One recommended way is the American Psychological Association method (known as the APA method). Lots of fields use this method, not just psychology. For data downloaded from the Internet, cite the Web page (see next page).

1. First create a text box for the source information. On the Insert menu, select Text.

2. The text box is dropped in the center of the map, which makes it difficult to work with. Drag the text box to the lower right corner so it's easier to see.

3. Right-click the text box and select Properties.

You can have a map source (your agency) as well as a data/ geography source (the census).

Type the following:

Map Source: Your organization, 2009.

Data Source: U.S. Census Bureau. (2009, January 1). 2009 TIGER/ Line Shapefiles for: Alabama. Retrieved November 1, 2009, from U.S. Census Bureau: http://www2.census.gov/ cgi-bin/shapefiles2009/ state-files?state=01. 5

1. Click Change Symbol and change the font from 10 to 12. Sources, as well as other map text, should generally not be smaller than 12-point font.

2. Click OK twice. Notice the citation is centered and lengthy.

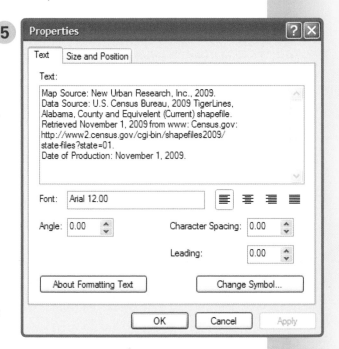

3. Right-click the Text box and select the align-left option.

4. Insert the cursor where you would like to break the line and press Enter. Do this four or five times. You can click Apply after each time to see how it looks. Once it looks reasonable, click OK.

5. Also now is a good time to type in the production date of the map.

6. Click OK.

Optional layout considerations

9 Insert north arrow

With north arrows, the simpler the better. Avoid overly ornate arrows.

1. From the Insert menu, select North Arrow.

2. The font is difficult to change on ArcGIS north-arrow templates. Select a north arrow with a font similar to fonts used in the map, ideally a sans serif one. ESRI North 6 is a good option.

3. Move the north arrow into the legend box and place it in the upper right corner. You will likely need to make it smaller so it fits in the box.

10 Map frame boundary

Whether your map should have a frame is up to you. If you decide to include a frame, stick with the default line type (simple) and width (0.5). To see what the map looks like without a frame, do the following:

1. While in Layout View, click the map frame (the whole map) to activate it. Blue dots should appear around the frame.

2. Right-click the map and select Properties.

3. Select the Frame tab.

4. Under Border, scroll to the top and select None.

5. Click OK.

MAP FLOW

The philosophy behind map flow is based on the principles of photography. The eye should be drawn to certain anchor points and move across the page. In this example, the eye moves in a Z pattern.

Avoid clustering all elements on one side of the map, which creates an unbalanced composition.

White space should be fairly equal on all sides of the map.

CHAPTER 3

Projecting shapefiles

Projections give shapefiles the correct shape, area, direction, and distance. Defining the datum, projection, and coordinate system for shapefiles will ensure that the geography is reflected properly, that distance (in the scale bar, for example) is recorded accurately, and that all layers are visible. Projections and datum can be complicated, but this chapter provides some general guidelines that will make your life easier.

This exercise is especially applicable if you download shapefiles from the Internet, ESRI Data & Maps DVDs, or your internal company network.

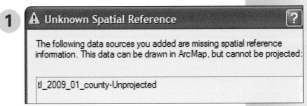

Exercise goal

Assign a datum and coordinate system to an unprojected shapefile. 1

Before we get started, you should know about six clues that will let you know if something is wrong with the projection of your file. If any one of these clues is present, you have a problem.

Exercise file locations

Let's make it easy on ourselves and use one of the shapefiles we downloaded in chapter 1. If you did not complete chapter 1, review the instructions on downloading a county shapefile from the U.S. Census Bureau, or simply use the file provided on the CD included with this book.

Chapter directions: Follow the exercise as it appears in this book
County shapefile used in chapter 1 (**tl_2009_01_county**).

Here is what you need to do to make this file viable for this exercise. Do not skip this step.

1. Create a new folder on your root C drive called **Ex3Projections**.
2. Copy the original chapter 1 county folder and its files. It will be named **tl_2009_01_county** unless you downloaded a different county.
3. Paste the files in the new folder called **Ex3Projections**.
4. Look at the files that make up the shapefile. For most users this can be accomplished by clicking the My Computer icon on your desktop and navigating to where your shapefile is saved. Here you should see a projection file. Delete this file. That deletes the projection associated with this file, thus making it unprojected.

CD: Use the CD included with this book
All files needed for this exercise are included on the book's CD. Files are organized by chapter.

Personal files: Use files you've gathered from other sources
Any shapefile will work for this exercise, although it will need to be modified. To modify the file, look at the files that make up the shapefile. For most users this can be accomplished by clicking the My Computer icon on your desktop and navigating to where your shapefile is saved. You should see a projection file. Delete the projection file. This deletes the projection associated with the file, making it unprojected.

Project a file

1 Add a shapefile

1. Open ArcGIS.

2. Add a shapefile to the ArcGIS window by clicking the Add Data icon. ➕ ▾

3. Either open Folder Connections or select the Connect to Folder icon and navigate to the modified shapefile for this exercise. 📂

4. Select the file to be projected and click Add.

5. You should get an error message for Unknown Spatial Reference.

6. Click OK. The shapefile will be added anyway.

2 Define datum

Projecting files is a two-step process. The first step is like selecting a one-size-fits-all projection for the shape-file. The second step is to fine-tune the projection to a smaller area (in this example Alabama). By fine-tuning it, the map will be more accurate. ❓

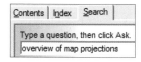

SIX CLUES YOU HAVE A PROJECTION PROBLEM

1. You get an error message, such as the one displayed on the first page of this exercise, when you first open the shapefile indicating "unknown spatial reference."

2. When you add a shapefile to the Data View, the file is not visible. This often happens when you are working with multiple layers from various sources.

3. You get an error message when you insert a scale bar.

4. The scale bar distance is inaccurate.

5. The map looks skewed.

6. On the Source tab in layer properties, the coordinate system is undefined.

DATUM

Geodetics is a part of the earth sciences field that works with the measurement and representation of the earth. To geodetics people, as well as all of us in any geoscience field, datum refers to the many little data points that were collected and analyzed to better understand and represent the curvature of the earth.

There are many types of datum since many people all over the world had the same idea to collect and analyze points to create a representative model of the earth. In North America the most commonly used is the datum NAD 83, which is based on 250,000 points, according to National Geodetic Survey.

For quick information about datum, do an Internet search for "National Geodetic Survey." For a very interesting book that brings geodetics to life, try *Longitude: The True Story of a Lone Genius Who Solved the Greatest Scientific Problem of His Time*, by Dava Sobel.

1. Click the ArcToolbox icon 📕 to open ArcToolbox.

2. Expand the Data Management Tool menu, then expand the Projections and Transformations menu.

3. Double-click Define Projection.

4. Use the first drop-down menu to select the shapefile you would like to project (the county shapefile) by clicking the little yellow file folder button.

5. Click the Coordinate System button. 🗔

6. Click the Select button.

7. Click the Geographic Coordinate Systems folder.

8. Select North America.

9. Double-click NAD 1983.prj.

10. Click OK twice.

   Your map may not look any different yet. It will after we complete the second step of this process.

COORDINATE SYSTEMS

A quick reference for coordinate systems for your local area can be found at **http://home.comcast.net/ ~rickking04/gis/spc.htm**.

3 Define the projected coordinate system

Next we need to fine-tune the coordinate system of this file to make it a better fit for Alabama.

1. Click the ArcToolbox icon ⧉ to open ArcToolbox.

2. Expand the Data Management Tool menu, then expand the Projections and Transformations menu and finally expand the Feature menu.

3. Double-click the first link, Project.

4. In the first field, click the Input Datasets or Feature Class arrow and select the county shapefile.

5. Input Coordinate System should automatically be filled in.

6. The Output Dataset or Features Class field represents where you would like to save what will be a newly projected shapefile. Navigate to where you would like to save it on your C drive. Give the file a new name.

   NOTE: If it's unclear how to navigate to the C drive, see step four, chapter 1.

7. Click the icon next to the Output Coordinate System field.

8. Click Select button.

9. Double-click the Projected Coordinate Systems folder.

10. Double-click the State Plane folder.

11. Double-click the NAD 1983 (US Feet) folder.

12. Select the appropriate state plane. In this exercise it is NAD 1983 (US Feet) State Plane Alabama East FIPS 0101.prj. If you are projecting another geography, select the appropriate geography here. To determine the appropriate geography, follow the Coordinate Systems link in the upper right corner of this page. In the future, if you don't know which one to select, choose any geography for your state for a close enough fit. For example, there are two possible options for Alabama representing East and West. We can only choose one, so select the first one from the list. If you were doing a smaller geography, a city, or single county boundary, it would be necessary to be more accurate and select the correct one. **2**

13. Click Add, then OK twice.

14. From the File menu, select New and open a new project. You must add your newly projected file to a new ArcGIS session before you are able to see any changes. Add the new projected file and notice how it looks slightly different.

15. In addition to looking at the geography, confirm that it worked correctly by right-clicking the layer and selecting Properties, then select the Source tab.

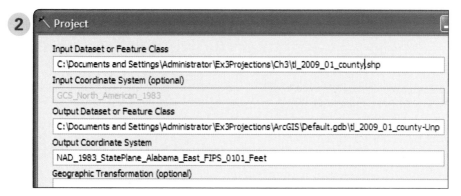

16. **Verify that the Projected Coordinate System is NAD_1983_StatePlane_Alabama_ East_FIPS_0101 and underneath that the Datum is D_North_American_1983.**

    Can you see how the new shapefile looks elongated compared to the original? The difference is not visually drastic (in some cases it will be); however, with the correctly projected shapefile, the scale bar is accurate. ③

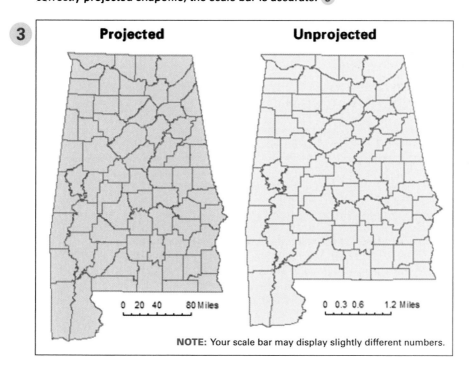

If you are displaying the entire United States, you may prefer to use North_ America_Albers_Equal_Area_ Conic as the projection so that the map looks more like the image on the bottom, with a more traditional projection with a curved northern boundary. **4**

To use North America Albers Equal Area Conic projection, follow step 2 in this chapter, with the following minor exceptions:

1. Double-click the Projected Coordinate Systems folder.

2. Double-click the Continental folder.

3. Double-click the North America folder.

4. Select the North America Albers Equal Area Conic.prj. (If you are projecting for Alaska or Hawaii, you should select the appropriate projection.)

U.S. SHAPEFILE

You can easily get a shapefile of the entire country from the U.S. Census Bureau Web site. See chapter 1 for instructions. Select State and Equivalent (Current) as the geography type. You'll need to download it and unzip it.

CHAPTER 4

Preparing data for ArcGIS

When shapefiles are downloaded from the census or other places, the usual situation is that the file only has a few columns of information and those columns are all related to spatial features of the shapefile. Boring! The real magic happens when you are able to display *your* data on a map. That's the good stuff, right?

Many people are familiar with how to use Microsoft Excel to some extent. It's a widely used program for data analysis and manipulation. If you have any familiarity with Excel, it will be much easier to prepare the data using that tool, instead of trying to do a lot of complicated manipulation in ArcGIS. Most commonly, spatial data is first derived and manipulated in Excel (or similar software programs). Then it is added to ArcGIS.

In this exercise you will download a data table from the U.S. Census Bureau Web site. The first few pages of this exercise walk you through the process of downloading a census table. Although we are using a census data table to illustrate data preparation for ArcGIS, many universally applicable techniques are built into this exercise. And if learning to prepare census data for ArcGIS is one of your goals, then this exercise is perfect for you. **1**

Exercise goal

Prepare data to use in ArcGIS.

The variable we intend to derive from the census is *senior population*, defined as individuals 65 years and older. A key objective of this exercise is to share useful Excel tricks that can enhance your work in GIS.

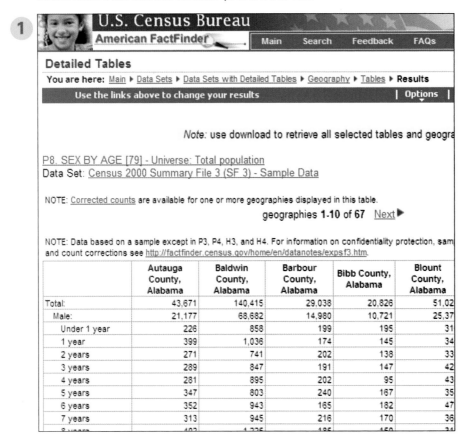

1

U.S. Census Bureau
American FactFinder

| | Main | Search | Feedback | FAQs |

Detailed Tables

You are here: Main ▶ Data Sets ▶ Data Sets with Detailed Tables ▶ Geography ▶ Tables ▶ **Results**

| Use the links above to change your results | | Options |

Note: use download to retrieve all selected tables and geogra

P8. SEX BY AGE [79] - Universe: Total population
Data Set: Census 2000 Summary File 3 (SF 3) - Sample Data

NOTE: Corrected counts are available for one or more geographies displayed in this table.

geographies 1-10 of 67   Next ▶

NOTE: Data based on a sample except in P3, P4, H3, and H4. For information on confidentiality protection, sam and count corrections see http://factfinder.census.gov/home/en/datanotes/expsf3.htm.

	Autauga County, Alabama	Baldwin County, Alabama	Barbour County, Alabama	Bibb County, Alabama	Blount County, Alabama
Total:	43,671	140,415	29,038	20,826	51,02
Male:	21,177	68,682	14,980	10,721	25,37
Under 1 year	226	858	199	195	31
1 year	399	1,036	174	145	34
2 years	271	741	202	138	33
3 years	289	847	191	147	42
4 years	281	895	202	95	43
5 years	347	803	240	167	35
6 years	352	943	165	182	47
7 years	313	945	216	170	36
8 years	402	1,226	185	150	34

Exercise file locations

In this chapter we will download demographic data from the census that will do the following:

- Provide raw, messy data to prepare for ArcGIS. In other words, real data.
- Create a file to be used for joining in chapter 5.
- Introduce you to the wealth of information freely available from the U.S. Census.
- Show you how to download census data to be used specifically with GIS.

This exercise uses Table P8. Sex by Age from the Census Summary File 3 (SF3) for Alabama Counties. If you are not working with census data, open any Excel spreadsheet you would like to map and apply the techniques listed below. **Note:** We will use this data in chapters 5 and 6.

Chapter directions: Follow the exercise as it appears in this book

All files for this exercise will be downloaded as a part of the exercise. We'll download census **Table P8. Sex by Age** from the Census Summary File 3 (SF3) for Alabama Counties.

We're using this geography because in chapter 1 we downloaded the Alabama counties shapefile. In this exercise we will get data for those geographies, and in chapter 5, we'll join the Excel spreadsheet that is derived from this exercise with the shapefile from chapter 1, thereby illustrating how to join an Excel spreadsheet to a shapefile.

CD: Use the CD included with this book

All files needed for this exercise are included on the book's CD. Files are organized by chapter.

Personal files: Use files you've gathered from other sources

You may select geography other than Alabama counties. You may also select a variable other than P8. Sex by Age, though this variable has been specifically selected for this exercise to show you key Excel operations.

Part 1: Downloading data from the U.S. Census

1 Get demographic data from the U.S. Census Bureau Web site

1. Go to **www.census.gov**.
2. Click the American FactFinder link on the left navigational bar.
3. On the left navigational bar, point to Data Sets and select Decennial Census.
4. Select the 2000 Summary File 3 (SF3). This is the most commonly used dataset from the census. It contains the most data, at the lowest level geography (block group).
5. To the right of the SF3 selection button is a list of options. Select Detailed Tables.

2 Select geography ②

1. Click the arrow on the geographic type box and select County.
2. In the next box, select the state Alabama.
3. Select All Counties from the list and click Add. This will populate the box at the bottom with all the counties in Alabama.
4. Once you have added the desired geography (captured in the box at the bottom), click Next.
5. You must now choose the data tables you would like to map. Select **P8. Sex by Age** and click Add. (For more experience, you can try using the Subject and Keyword searches to get familiar with some of the variables.) Then click Show Result.

   This will yield an HTML page with all your data on it.

Downloading data for mapping

3 Save data

1. To save the data table, from the Print/Download menu at the top, select Download. ③
2. Several options are available for saving. To merge geography files with data tables later on, we must have a unique identifier column. Under the Database compatible download, select the Microsoft Excel option to provide this column. ④

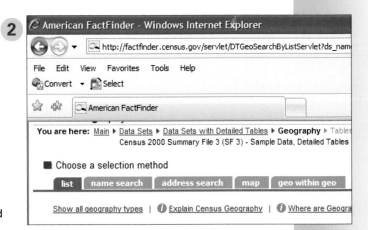

U.S. CENSUS GEOGRAPHIES CAN BE CONFUSING

The U.S. Census site allows you to select shapefiles and tabular data for many different types of geography (tracts, counties, the entire nation, states, etc.).

Here is a quick reference of the most widely used geographies:

Nation: This is for the United States as a whole. If you select this geography and then, for example, Population as a data variable, the result will be one number, the population of the entire United States.

State: Allows you to select one state, multiple states, or all states.

County: Allows you to select one county, multiple counties, or all counties for the entire United States. (Although, that would be a lot of counties!)

Place: Represents city boundaries, plus Census Designated Places.

Census tract: Tracts are the most popular subcounty geography. They are fixed in population between 1,000 and 8,000 people. Census tracts average about 4,000 people, although this varies.

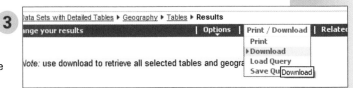

POP-UP BLOCKER ALERT

If your computer has a pop-up blocker, hold down the Ctrl key to force it to allow the dialog box.

3. While holding the Ctrl key, click OK. This will circumvent the pop-up blocker. Please do not skip this step.

4. Keep holding the Ctrl key until you are prompted to save the zipped file.

5. Navigate to where you would like to save your file (preferably on the root C drive), click Save, then Close.

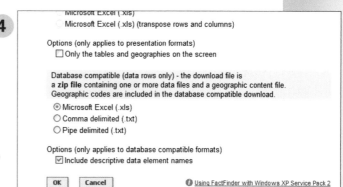

4 Unzip the file

1. Navigate to your file, find the zipped file you just saved and right-click the file. Follow the prompts to unzip. Each unzipping program is different but generally you're looking for something that says Extract to here or Extract all.

2. Once you have unzipped the output file, you should have four new files. The file that contains the data is **dt_dec_2000_sf3_u_data1.xls**.

Part 2: Prepping data for ArcGIS

The purpose of preparing data in Excel first, instead of in ArcGIS, is to simplify the data, focus on what you want to join (and eventually thematically map), and create the optimal spreadsheet to import to ArcGIS.

When downloading census data, you often get numerous columns of irrelevant infor-

PROBLEM ALERT

When you open the **dt_dec_2000_sf3_u_data1.xls** file, you should see several rows of data. If you only see twelve rows of data, you most likely did not hold down the Ctrl key properly (see step 3).

You must hold the Ctrl key down throughout the *entire* save. For an easy fix, delete the output file you just downloaded.

Then go back to the census site and reselect the geography (by clicking the geography link at the top of the page).

Once you have reselected all geographies, redownload the file while holding down the Ctrl key *throughout the entire save*.

The new download will display many rows of data in your spreadsheet.

ESSENTIAL FIPS CODES

Be absolutely sure to keep the *second* geography identifier column that contains the FIPS code (Federal Information Processing Standards code). This code is how you will link your spreadsheet to a shapefile.

FIPS codes provide a unique ID for every parcel of land in the United States. The eleven-digit code in column B represents the state code (31 for Nebraska), the county code (05500 for Douglas County), and finally the tract number code (200 for tract 2).

mation. It is helpful to decide early on exactly what you would ultimately like to display on your map. In chapters 5 and 6, we will join and create a thematic map using senior population. We don't need to bring all of this data into ArcGIS since we only need a small portion of it. Also, you'll notice a column for senior population is not given in the downloaded census data. You'll need to derive this data from the given census data. The steps below are helpful when working with census data and many other kinds of data. These techniques will prove useful over and over again.

5 Delete unnecessary columns and rows

1. Delete the first column Geography Identifier and the Geographic Summary Level column because they are unnecessary. Be careful that you do not delete the second Geography Identifier column (see sidebar about the FIPS codes).

2. Delete row 1 so you are left with the text headings instead of numeric headings.

3. Delete all columns that are not relevant to the variable you intend to map. In this exercise you are mapping percentage of the population 65 years and older also referred to as the Senior Population, so delete all age columns less than 65.

> ### HEADER ROWS
> If you bring a spreadsheet with two rows for a header into ArcGIS, the program gets confused. You can open it in ArcGIS, but you cannot join a spreadsheet with two header rows to a map.

> ### ISOLATING RELEVANT DATA
> At this point, we need to be hyperfocused on isolating the relevant data. This approach should be applied to prepping all spreadsheet data to bring into ArcGIS.

NOTE: Keep the Total Population column because you'll need it later.

Once you do this, you should be left with fifteen columns: Total Population plus all age groups 65 and greater for both men and women and two columns of geography information.

6 Clean up county column (optional)

1. You might notice that column B (with the column header Geography) looks a little funny. If so, you need to widen the Geography column so you can see all the text. *Make the column as wide as necessary.*

2. The Geography column text reads Autauga County, Alabama, but for labeling purposes later, it would be helpful to just have the county name with no additional text. To accomplish this, use Excel's Find & Replace function. Highlight the Geography column, select Find & Replace (Excel 2000) from the Edit menu or Find & Select (Excel 2007) on the Home tab (2007).

3. In the Find What Field, type **County, Alabama**. You may not be able to see it here, but there is a space before the word County. **5**

5

Find	Replace
Fin_d what:	County, Alabama ⌄
Re_place with:	⌄

Options >>

4. You will replace it with no text, so don't type anything in the Replace with field. Click the Replace All button. It should end up looking like the image here. 6

6

	A	B	
	Geography Identifier	Geography	Tota
1			
2	01001	Autauga	
3	01003	Baldwin	
4	01005	Barbour	
5	01007	Bibb	
6	01009	Blount	
7	01011	Bullock	

B1 ▼ f_x Geography

> **GOOD TO KNOW**
>
> This technique is often used with census tract numbers to isolate the number for joining purposes later.

7 Derive data

You can derive data in Excel by using the sum, divide, and take percentages functions. These steps are integral to working with data in ArcGIS since we ultimately want to map the senior population, which is defined as 65 years and older. Since these numbers are not simply given with the downloaded data, we need to complete a series of steps to get the right information. These steps include the following:

1. Sum the twelve columns that contain the numbers for the population 65 years and older to get the total sum of people 65 plus.

    The sum formula in Excel is **=sum(range that you want to sum)** `=SUM(D2:O2)`

    Here is how to sum:

2. Go to the first empty column at the far right of the spreadsheet and type the column name **Seniors**.

    a. Click the empty cell in row 2.

    b. In that cell, type **=sum(**. 7

7

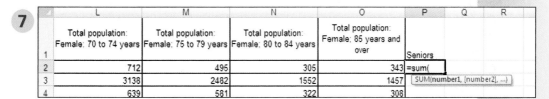

	L	M	N	O	P	Q	R
	Total population: Female; 70 to 74 years	Total population: Female; 75 to 79 years	Total population: Female; 80 to 84 years	Total population: Female; 85 years and over			
1					Seniors		
2	712	495	305	343	=sum(		
3	3138	2482	1552	1457	SUM(number1, [number2], ...)		
4	639	581	322	308			

c. Using the left arrow key, scroll to column D (Total population: Male; 65 and 66 years), hold down the Shift key, and use the right arrow key to continue to highlight the range you would like to sum. The range starts with column D and ends with the last column (column O). Click the Enter key (this does the sum). The number you should get is 4442, the number of seniors in Autauga County, Alabama.

d. To carry this formula all the way down the column, you could drag the formula down, but here's a cool Excel trick. In the lower right corner of the cell, you can see a little black dot. If you hover over the dot with your cursor, you will see a little black plus sign. When you see that plus sign, double-click and the program will copy the formula down the entire column. It will copy it down only as far as it needs to.

Next, you'll want to derive the percentage of the population that is 65 plus by county. To do this, divide the senior population by the total population.

To divide in Excel, type the following formula: **= (numerator/denominator)**.

Here is how you derive a percentage:

Scroll to the first empty column to the right and at the end of your spreadsheet and name the column **Percent**. This column should be next to the Seniors column.

   e. Click the empty cell in row 2.

   f. In that cell, type an equal sign = (this initiates the calculation).

   g. Using your mouse, click the number in P2 (in the row where your data begins).

   h. Type / (the division symbol).

   i. Using your mouse, scroll over to the beginning of the spreadsheet and click the cell C2 (this is the total population). **8**

   j. Click Enter.

   k. Scroll back to the end of your spreadsheet and verify that the percentage was calculated correctly. It should be 0.101715 (although you may have more or fewer digits showing).

8

	N	O	P	Q
1	Total population: Female; 80 to 84 years	Total population: Female; 85 years and over	Seniors	Percent
2	305	343	4442	=P2/C2
3	1552	1457	21674	
4	322	308	3915	

   l. To carry this formula all the way down the column, hold your cursor over the little black dot in the lower right corner of the cell. When you see the little plus sign, double-click and the program will copy the formula down the entire column. (At this point, the numbers are still fractions but will be converted to actual percentages in chapter 6.)

GOOD TO KNOW

Do not reformat the percentage. It's better to leave percentages as a General Format in Excel, to be changed later in ArcGIS. If the percentage column contains all zeros when opened in ArcGIS, go back to the Excel file, recalculate the field and leave it in General Format.

8 Turn formulas into real numbers

The two derived columns (Seniors and Percent) contain formulas, not real numbers. One thing you will want to do is simplify your spreadsheet by deleting all those columns for men and women because we now have that data summed up in one column. Your spreadsheet will be much easier to work with in ArcGIS if it's streamlined. To delete these columns of data and not mess up the Seniors and Percent columns, you must first perform a series of steps that will turn the formulas into real numbers.

To accomplish this, do the following:

1. Highlight the two new columns Seniors and Percent. You can highlight the column by clicking the letter at the top of that column (in this case P) and then dragging with your mouse to the right to include the Q column as well.

2. Right-click inside the blue highlighted area and select Copy.

3. Right-click again inside of the blue highlighted area and select Paste Special.

4. Select the button Values and the OK button, then click Enter. This will leave only values in the columns instead of formulas.

5. To clean up the spreadsheet and make it more manageable, we'll delete the twelve age columns and only keep the following columns:

   | Geography Identifier | Geography | Total Population | Seniors | Percent

9 Change column headings

In ArcGIS software versions older than 9.3, you can only use ten characters or fewer for column headings (otherwise, the spreadsheet will not join properly). Also, in older versions you cannot have spaces in the column name (or file name) or any nonalphabet characters such as periods or commas. Even though these issues have been fixed in ArcGIS 10, it's still a good habit to shorten names and take out spaces.

1. Geography Identifier: rename **JoinID** (since this is the column that we'll use in the next chapter when joining).

2. Geography: rename **County**.

3. Total Population: rename **Population**.

4. The other two columns are **Seniors** and **Percent**.

   The final spreadsheet should look like this: **9**

	A	B	C	D	E
	JoinID	County	Population	Seniors	Percent
1				Seniors	Percent
2	01001	Autauga	43671	4442	0.101715
3	01003	Baldwin	140415	21674	0.154357
4	01005	Barbour	29038	3915	0.134823
5	01007	Bibb	20826	2414	0.115913
6	01009	Blount	51024	6462	0.126646
7	01011	Bullock	11714	1513	0.129162
8	01013	Butler	21399	3516	0.164307
9	01015	Calhoun	112249	15825	0.140981
10	01017	Chambers	36583	5889	0.160976
11	01019	Cherokee	23988	3835	0.159872
12	01021	Chilton	39593	5081	0.128331
13	01023	Choctaw	15922	2294	0.144077
14	01025	Clarke	27867	3742	0.134281
15	01027	Clay	14254	2378	0.16683
16	01029	Cleburne	14123	1947	0.13786
17	01031	Coffee	43615	6229	0.142818
18	01033	Colbert	54984	8478	0.15419
19	01035	Conecuh	14089	2229	0.158209
20	01037	Coosa	12202	1734	0.142108
21	01039	Covington	37631	6720	0.178576
22	01041	Crenshaw	13665	2366	0.173143
23	01043	Cullman	77483	11321	0.146109

L23

AGE

10 Rename the worksheet and save

1. Naming Excel worksheets helps you stay organized. To do this, in the lower left corner double-click the Sheet0 tab and type **AGE**. You are able to bring in multiple worksheets from the same Excel Workbook.

2. Now save your Excel spreadsheet and name your new spreadsheet **Age.xls** (or **.xlsx** depending on the Excel version).

EXCEL AND OLDER VERSIONS OF ARCGIS

Older versions of ArcGIS cannot read Excel 2007 files (with the file extension .xlsx). Also, if you are using an older version of ArcGIS, be sure to shorten column headings and save your file as a database file (.dbf), instead of Excel. You can always bring a database file into ArcGIS, and often they are less error prone than Excel spreadsheets.

SPSS THINGS TO KNOW

Any blanks in SPSS files will be imported as zeros in ArcGIS. Often blanks and zeros are very different things.

CHAPTER 5

Joining data to maps

One of the most frequently used GIS skills involves connecting an Excel spreadsheet of data to a shapefile. Often, the purpose of joining data to a map is to visually display the distribution of a dataset through a thematic map (covered in the next chapter). Joining your own data to a shapefile can be extremely useful.

Exercise goal

Join the senior population spreadsheet created in the last exercise to a county shapefile (downloaded in exercise 1). **1**

Exercise file locations

Chapter directions: Follow the exercise as it appears in this book

Files for this exercise were created and used in chapters 1 and 4.

The following files are used in this exercise:

- **Age.xlsx** (created in chapter 4)
- **tl_2008_01_county.shp** (a shapefile of Alabama counties that was downloaded in chapter 1).

Note: The file derived in this chapter will be used again in chapters 6 and 7.

CD: Use the CD included with this book

All files needed for this exercise are included on the book's CD. Files are organized by chapter.

Personal files: Use files you've gathered from other sources

This exercise assumes you have the following:

- A shapefile with an attributes table that includes several columns of data.
- An Excel spreadsheet with several columns of data that you would like to join to a shapefile.
- Both files must have overlapping data. Often each file has one identical column of geographic data, such as the name of the county or census tracts. This is the column that will be used to join the two files together.

1 Add two files to join

1. Open ArcGIS.

2. To add a shapefile to the ArcGIS window, click the Add Data icon ⊕ ▾ .

3. Either open Folder Connections or select the Connect to Folder icon and navigate to **tl_2009_01_county**, which was downloaded in chapter 1.

4. Use the Add Data icon again to add the Excel file. In the previous exercise, you named this worksheet **Age.xlsx**. To add the Excel worksheet, navigate to the file and double-click the file name. Double-click **AGE$**. If you did not change the name of the worksheet in the last exercise, the existing worksheet will be called **Sheet0$**. Worksheets are denoted with $ in the name. **2**

5. Check to make sure the data is correct. The **AGE$** data table should now appear in the table of contents. To view the data table, right-click the data table and select Open. Review the data to

2

	FID	Shape	STATEFP	COUNTYFP	COUNTYNS	CNTYIDFP	NAME	NAMELSAD	LSAD	CLASSFP
▶	0	Polygon	01	113	00161583	01113	Russell	Russell County	06	H1
	1	Polygon	01	067	00161559	01067	Henry	Henry County	06	H1
	2	Polygon	01	029	00161540	01029	Cleburne	Cleburne County	06	H1
	3	Polygon	01	037	00161544	01037	Coosa	Coosa County	06	H1
	4	Polygon	01	093	00161573	01093	Marion	Marion County	06	H1
	5	Polygon	01	099	00161576	01099	Monroe	Monroe County	06	H1
	6	Polygon	01	107	00161580	01107	Pickens	Pickens County	06	H1
	7	Polygon	01	129	00161590	01129	Washington	Washington County	06	H1
	8	Polygon	01	007	00161529	01007	Bibb	Bibb County	06	H1
	9	Polygon	01	013	00161532	01013	Butler	Butler County	06	H1
	10	Polygon	01	023	00161537	01023	Choctaw	Choctaw County	06	H1
	11	Polygon	01	057	00161554	01057	Fayette	Fayette County	06	H1
	12	Polygon	01	071	00161561	01071	Jackson	Jackson County	06	H1
	13	Polygon	01	133	00161592	01133	Winston	Winston County	06	H1
	14	Polygon	01	003	00161527	01003	Baldwin	Baldwin County	06	H1
	15	Polygon	01	001	00161526	01001	Autauga	Autauga County	06	H1
	16	Polygon	01	005	00161528	01005	Barbour	Barbour County	06	H1
	17	Polygon	01	015	00161533	01015	Calhoun	Calhoun County	06	H1
	18	Polygon	01	027	00161539	01027	Clay	Clay County	06	H1
	19	Polygon	01	063	00161557	01063	Greene	Greene County	06	H1
	20	Polygon	01	033	00161542	01033	Colbert	Colbert County	06	H1
	21	Polygon	01	009	00161530	01009	Blount	Blount County	06	H1

L23

	A	B	C	D	E
1	JoinID	County	Population	Seniors	Percent
2	01001	Autauga	43671	4442	0.101715
3	01003	Baldwin	140415	21674	0.154357
4	01005	Barbour	29038	3915	0.134823
5	01007	Bibb	20826	2414	0.115913
6	01009	Blount	51024	6462	0.126646
7	01011	Bullock	11714	1513	0.129162
8	01013	Butler	21399	3516	0.164307
9	01015	Calhoun	112249	15825	0.140981
10	01017	Chambers	36583	5889	0.160976
11	01019	Cherokee	23988	3835	0.159872
12	01021	Chilton	39593	5081	0.128331
13	01023	Choctaw	15922	2294	0.144077
14	01025	Clarke	27867	3742	0.134281
15	01027	Clay	14254	2378	0.16683
16	01029	Cleburne	14123	1947	0.13786
17	01031	Coffee	43615	6229	0.142818

make sure it looks as you would expect it to look.

To join data to maps, we must link two columns that have overlapping data, one column from the data table and its comparable column in the map layer. In this case, the columns will be identical.

PROBLEM ALERT

If you are using a version of ArcGIS that is older than 9.3, you cannot have more than ten characters for column names. This must be fixed ahead of time. The two files will not join properly if you have more than ten characters, spaces in column names, periods, or any character that is not a letter (no numbers).

2 Join the data table to the map

1. Identify the two columns you will use for joining by opening the attributes table for each. The attributes table for Age may already be open from the previous step. Right-click the county shapefile in the table of contents, and select Open Attributes Table. Notice that two tabs are now open at the bottom of the table.

   NOTE: Working with attribute tables is covered in detail in chapter seven.

2. Click each one of the tabs in the lower left corner, scan the data in each, and find the two columns that match. The column names do not have to be the same, but the content of the columns does. In this example, the column name in the shapefile attribute table is CNTYIDFP and the column name in the spreadsheet is JoinID.

3. In the table of contents, right-click the shapefile name (not the data table).

4. Select Joins and Relates from the menu.

5. In the field Choose the field in this layer that the join will be based on, select the appropriate column heading, in this case CNTYIDFP.

6. In the field Choose the table to join to this layer, select AGE$.

7. In the field Choose the field in the table to base the join on, select the appropriate column heading, in this case JoinID.

8. Select the Keep Only Matching Records option.

9. Click OK. ③

③

CNTYIDFP	JoinID
01113	01001
01067	01003
01029	01005
01037	01007
01093	01009
01099	01011
01107	01013
01129	01015
01007	01017
01013	01019
01023	01021
01057	01023
01071	01025

3 Verify that the join worked correctly

1. Right-click the shapefile and select Open Attributes Table.

2. Scroll to the far right to see if the data from the Excel spreadsheet has been appended to the end of the attribute table. You should not see any error messages or null values.

3. Double-check the number of records in the Age$ tab by right-clicking the file and selecting Open. In the lower right corner, the number of records is listed (in this case, 67). Now check the number of records in the newly joined shapefile—it should be the same number. If it is not, then the two columns are not identical and must be corrected.

4. Close the table window by clicking the red x in the upper right hand corner.

4 Create a new shapefile

When files are joined, it's a temporary join. To permanently join these files, you must create a new shapefile with the merged files. To do this, do the following:

1. In the table of contents, right-click the shapefile name, select Data, and then select Export Data.

2. Click the browse folder icon to name your new shapefile and to save it on your computer. Name it **AgeJoined**. Never use spaces in file names as it will generate all sorts of problems.

3. When asked if you want to add the exported data to the map as a layer, click Yes.

4. The original Excel file and shapefile are no longer needed. To remove them, right-click the AGE$ file and select Remove. Then right-click the original county file and select Remove. Congratulations! You now have a permanently joined shapefile (AgeJoined) that contains data about the senior population. We will use this file in the next exercise.

CHAPTER 6

Creating a thematic map

Mapping thematically enables you to show the distribution of data across geography. It's one of the most frequently used GIS tools.

Exercise goal

Create a thematic map with graduated colors to show the distribution of the senior population (defined as people over age 65) in Alabama. **1**

Exercise file locations

Chapter directions: Follow the exercise as it appears in this book

This exercise uses a file created in chapter 5 called **AgeJoined.shp**.

CD: Use the CD included with this book

All files needed for this exercise are included on the book's CD. Files are organized by chapter.

Personal files: Use files you've gathered from other sources

For this exercise you need a shapefile, such as population by state, which lends itself to being thematically mapped.

Note: You will use the map you create in this exercise again in chapter 19, "Publishing maps."

1 Add a shapefile

1. Open ArcGIS.

2. To add a shapefile, click the Add Data icon ✚ ▾ .

3. Either open Folder Connections or select the Connect to Folder icon and navigate to **AgeJoined. shp**, created in chapter 5.

4. (Optional) To refresh your memory about what data is associated with this map, open the attributes table by right-clicking the layer name in the table of contents. Select Open Attributes Table and scroll to the far right to see the appended data.

2 Create a thematic map

1. In the table of contents, right-click the layer name and select Properties. Select the Symbology tab.

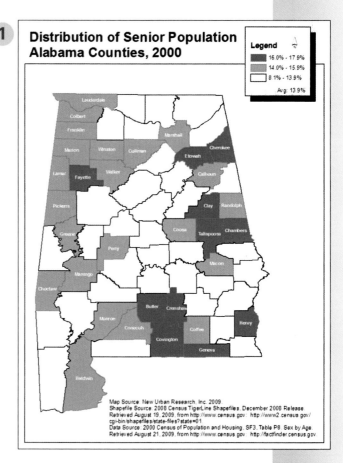

1

Distribution of Senior Population Alabama Counties, 2000

Legend

■ 16.0% - 17.9%
▨ 14.0% - 15.9%
□ 8.1% - 13.9%

Avg: 13.9%

Map Source: New Urban Research, Inc. 2009.
Shapefile Source: 2008 Census TigerLine Shapefiles, December 2008 Release.
Retrieved August 19, 2009, from http://www.census.gov: http://www2.census.gov/
cgi-bin/shapefiles/state-files?state=01
Data Source: 2000 Census of Population and Housing, SF3, Table P8, Sex by Age.
Retrieved August 21, 2009, from http://www.census.gov: http://factfinder.census.gov

2. On the left side of the box, select Quantities and Graduated Colors.

3. In the Value field, select the column of data you would like to see displayed on the map, in this case Percent.

4. Select a color ramp for your data or leave the default. Monochromatic blue is a good choice for thematic maps because it looks professional and readers can easily distinguish between the shades of blue.

5. Click OK to see the colors displaying the data on your map.

3 Fix the percent column

To change the percent column to show percentages instead of fractions, follow these steps:

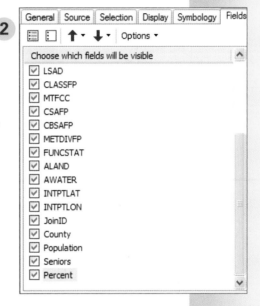

1. In the table of contents right-click AgeJoined and select Properties, then select the Fields tab (center). ②

2. On the left, scroll down to select the Percent field.

3. On the right, click the ellipsis box (to the right of Number Format).

4. Select Percentage as the category.

5. Select the second option: The number represents a fraction. Adjust it to show a percentage.

6. Click Numeric Options. [Numeric Options...]

7. The number of decimal places should be changed to one instead of the default six.

8. Click OK until all the dialog boxes are closed. It will not immediately be apparent that anything has changed.

9. In the table of contents, right-click AgeJoined and select Open Attribute Table. In the table, verify that the percentages are now displayed properly (scroll all the way to the far right).

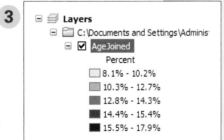

10. To get the new numbers to display properly, in the table of contents, right-click the layer name and select Properties. On the Symbology tab, reselect the Percent column. This refreshes the view, and now the correct form of the percentage displays. ③

4 Change the number of classes

Classes or interval ranges are the data ranges that appear on a map's legend. The default is five. This particular data lends itself to three intervals because the width of the intervals is so small. Normally, four intervals are preferable for a general, nontechnical audience.

1. Right-click the shapefile name in the table of contents, and select Properties and the Symbology tab.

2. Change the number of classes from five to three in the Classes box.

3. Click Classify and notice how the data is distributed along a normal curve. The method for breaking the legend is known as *natural breaks*, which is the default method.

4. After you see the data distribution, click OK to get out of the information box and click OK again to create three intervals. Now have a look at your map.

Two areas that can cause confusion when looking at thematic maps are the legend, specifically your legend break points, and color scheme. We'll use a technique on the legend, called the baseline method, to help readers easily see the hot spots on your map.

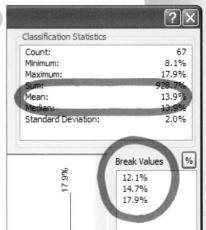

5 Change legend break points

The natural breaks method for legend creation is suitable for informal mapping. However, for a more sophisticated approach you can manually break the legend. And an even more sophisticated way than to simply randomly decide where to break it, is to use the average for your dataset as the first break point (thereby establishing a baseline for normalcy).

1. In the table of contents, right-click the shapefile name, and select Properties and the Symbology tab.

2. Click Classify to change the break values.

3. Look in the statistics box in the upper right corner to determine the mean. Write down that number. In this example, it's 13.9 percent. 4

BASELINE METHOD

To use the baseline method, the mean can be used for most data types. Exceptions would be income or housing values, in which case the median would be used.

4. To change the first break point to **13.9** percent, manually type over the given numbers in the Break Values section. The data distribution here only goes to 17.9 percent, so the break points after the 13.9 percent will have to be very small increments. You never change the last break point, as this is the natural end of your data. Change the break points to **15.9** and **17.9**, click OK twice and have a look at the map. Notice how you can more easily see a pattern emerging from the data.

COLOR

Most map readers intuitively understand that dark colors represent higher values and light colors represent lower values. When choosing your colors, remember that a dark-to-light monochromatic color scheme will generally be understood by most readers.

6 Change colors (or symbology in ArcGIS) ⑦

Type a question, then click Ask.
symbolizing data

Now that you've applied color to your map, a pattern may be emerging. You can make the pattern even clearer. De-emphasize less interesting areas by not shading a color, and emphasize areas of interest by applying color. In general, the less color you have on your map, the easier it will be to read and the better you can emphasize areas of interest. Without exception, using white to denote the lesser values in the map will make a pattern easier to see and make your map more professional looking.

Let's change the colors to emphasize areas that have a high proportion of the particular variable you're highlighting.

1. In the table of contents, right-click the shapefile name, and select Properties and the Symbology tab.

2. Change the lowest range (usually the first row) to white. You can simply double-click the color and change the fill color. Next, change the second range (class) to a light blue (assuming blue is your color palette). This symbology choice allows you to de-emphasize the low data values, while emphasizing the high data values with color. See the map on the first page of this exercise for an example of how your map should look.

7 Reverse the color order on the legend

When you want a map to emphasize the concentration of a variable (versus the lack of a variable), construct the legend so the highest values are at the top.

1. In the table of contents, right-click the shapefile name, and select Properties and the Symbology tab.

2. Left-click the Range column header and choose Reverse Sorting.

3. Left-click the Symbol column header and choose Flip symbols.

8 Label counties

Turn on the labels for counties and shade each white and bold. See the map at the beginning of this exercise to see the effect this will have. (How to turn labels on is covered in chapter 1, steps 6 and 7).

9 Create a layout

On the View menu, select Layout View. Create a layout that includes a legend, title, source, and north arrow. Scale bars are often left off thematic maps since distance is not relevant. Information on how to create a layout is provided in chapter 2.

10 Save the project

1. On the File menu, select Save As.
2. Navigate to where you would like to save on your C drive.
3. Type **Seniors_Thematic.mxd**.

    NOTE: If you plan on doing the exercise in chapter 19, "Publishing maps," you will use this saved project again.

11 Save as a layer package

A layer package preserves the color shading of the file as well as the attributes. Sometimes it's helpful to save layer packages for easy transfer of the layer and data.

1. In the table of contents, right-click the shapefile name.
2. Select Save as Layer Package.
3. Navigate to where you would like to save on your C drive.
4. Give it a new name (notice that the file contains the .lpk extension).
5. Click the Save button.
6. Close ArcGIS.
7. If you want to see the layer again, open ArcGIS.
8. Use the Add Data icon to add the layer file.

CLEANING UP THE LEGEND

Often you can make the legend clearer by removing layer names and column headings. In this example, the legend doesn't really need the text "AgeJoined" and "Percent."

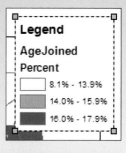

To remove this text, click the legend to activate it, right-click the legend, and select Properties. Click the Items tab, under Legend Items: right-click AgeJoined-Percent, and select Properties.

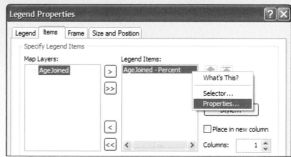

Make sure the Show Layer Name and Show Heading boxes are not selected, then click OK twice.

Occasionally, you may encounter an issue with the legend in which you've already selected white as the background color, but you can still see through parts of the legend around the edge. The fix for this is to delete the legend and reinsert it, again selecting white as the background color.

CHAPTER 7

Working with attribute tables

Manipulating data tables (or attribute tables in ArcGIS) includes such things as adding and deleting columns in a table and editing values or performing calculations. You will do this frequently when working with GIS. Although the bulk of data manipulation is best achieved outside of ArcGIS, it is important to learn the fundamentals of data manipulation within ArcGIS. The ability to edit data will greatly strengthen analysis and improve maps.

Exercise goal

Learn common ways to manipulate data within an attribute table.

Table

AgeJoined

FID	Shape *	STATEFP	COUNTYFP	COUNTYNS	CNTYIDFP	NAME	NAMELSAD	LSAD	C
0	Polygon	01	113	00161583	01113	Russell	Russell County	06	H1
1	Polygon	01	067	00161559	01067	Henry	Henry County	06	H1
2	Polygon	01	029	00161540	01029	Cleburne	Cleburne County	06	H1
3	Polygon	01	037	00161544	01037	Coosa	Coosa County	06	H1
4	Polygon	01	093	00161573	01093	Marion	Marion County	06	H1
5	Polygon	01	099	00161576	01099	Monroe	Monroe County	06	H1

Exercise file locations

Chapter directions: Follow the exercise as it appears in this book

This exercise uses the same AgeJoined.shp file used in chapters 5 and 6.

CD: Use the CD included with this book

All files needed for this exercise are included on the book's CD. Files are organized by chapter.

Personal files: Use files you've gathered from other sources

In order to complete this exercise, you will need a shapefile with an attribute table that has several columns of data that you can edit.

1 Add a shapefile and open the attribute table

1. Open ArcGIS.

2. To add a shapefile to the ArcGIS window, click the Add Data icon ✛ ▾ .

3. Either open Folder Connections or select the Connect to Folder icon and navigate to **AgeJoined.shp**.

4. To open the attribute table, right-click the shapefile in the table of contents and select Open Attributes Table. You should see your data in spreadsheet format.

NUMERIC COLUMN TYPES CAN BE CONFUSING

Generally, you should use float or double (also known as a double float) as the column type.

The double float is beneficial because you can have the maximum number of digits for the number, with no rounding (after 15 digits, plain old floats start to round, but doubles don't).

2 Add a column to the attributes table

1. Click the drop-down arrow for the Table Options on the Table toolbar (the first icon) and select Add Field. 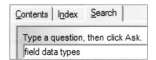 2

2. Type the name of the new column. Let's call it **NewColumn**. For the Type, select Float. Float is a very flexible column type, so it is often used. Float (as well as Double) is useful because the number can be in decimal format or a whole number. ?

Contents	Index	Search

Type a question, then click Ask.

field data types

3. You can leave the Precision and Scale as zeros for the maximum digits.
4. Click OK.
5. The column will be added to the end of attributes table. You can move the column by selecting the column and dragging it to where you want it in the table.

3 Edit existing data in an attribute table

If you try to edit any of the data in the table, you'll notice the values cannot be changed. You must first make the table editable.

1. Click the Editor Toolbar button.

In ArcGIS 10 this button should already be visible as it's a part of the default ArcGIS interface. In older versions of the software, you must go to the View menu, select Toolbars, and then select Editor. The icon will look the same.

2. Once you click the Editing icon, a new Editor toolbar should be visible. Dock the toolbar.

3. Click Editor and select Start Editing. Notice that the top row of the attributes table turns white, which means that the table is now editable.

GOOD TO KNOW

If multiple shapefiles are open, the Editor tool needs to know which files to edit. In this case, another dialog box would display asking you to start editing, and it would be necessary to pick which layer or which geodatabase with several files to edit. Here we are only working with one file, so it doesn't ask this.

4. Now let's edit some data. Go to any of your data values and type over the text to change the number to something else. For example, change the percentage of seniors in Cherokee County from 16 percent to **0** percent.

5. After you are finished editing data, click Editor again, and then select Stop Editing.

6. Save Edits when prompted.

7. Close the attributes table.

4 Edit data outside of the attributes table on a polygon-by-polygon basis

You can also edit data within what appears to be the information box instead of doing it within the attributes table. One advantage of this method is that you can click individual polygons and edit data for that polygon. To do this, do the following:

1. On the Editing toolbar, click the Editor button and select Start Editing.

2. On the Editing toolbar, click the Attributes icon.

3. On the map, click any county. Notice that a new window opens on the right with all the information from the underlying attributes table.

4. Click any field to edit the values.

5. Select Editor, Save Edits, and Stop Editing.

5 Make calculations

Calculations can be performed in or out of an editing session. Let's pretend we have not calculated the percentage of senior population. For the purpose of this exercise, let's recalculate the percentage of senior population.

1. Right-click the NewColumn header.

2. Select Field Calculator.

3. The formula to calculate the percentage of seniors is "seniors divided by population." Double-click Seniors from the list of variables, type / (the division symbol), and then double-click Population from the variables list. The formula is autofilled as you go (see right). Click OK. ③

4. The software calculates the percentage, but the column represents a fraction; we would like it to display as a percentage. You will need to reformat the column.

5. Right-click the column heading and select Properties. Click the ellipsis to change the column type.

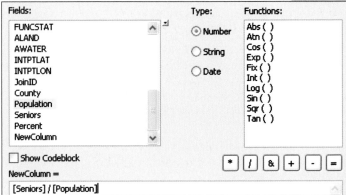

6. Select Percentage (on the left) then select the second option (on the right), The number represents a fraction. Adjust it to show as a percentage.

7. Click Numeric Options. Change the decimal places from 6 to **1**.

8. Click OK three times.

9. Now let's sort the column to see which counties had the highest rates of seniors. Right-click the column heading and select Sort Descending (Covington).

10. (Optional) To save column reformatting changes, you must save the project.

6 Delete columns

You may notice several prebuilt columns when you download your shapefiles from the census. Some columns are useful for understanding what geography is within the shapefile, but many of these columns can be deleted as they are rarely used. There are two ways to delete columns. The method you use depends on how many columns of data you want to delete. If it's only one or a few columns, you would use one method. However, for several columns, a different method must be employed.

To delete a few columns of data, do the following:

1. Highlight the column LSAD by clicking the column header.

2. Right-click and select Delete Field.

3. Select Yes to delete the column LSAD.

4. Now highlight the six columns: ClassFP, CSAFP, CBSAFP, METDIVFP, FUNCSTAT, JoinID. Right-click and select Delete Field. Notice that only the first field was deleted. To delete multiple columns, see the next step.

To delete multiple columns of data, do the following:

5. Close the attribute table.

6. In the table of contents, right-click the shapefile name, select Properties, and then click the Fields tab (center).

7. Click the Turn All Fields Off button ⬚ and notice that the check boxes next to all the columns become unchecked.

8. Now, select only those columns that you want to keep. The essential columns to keep are CNTYIDFP, County, and NewColumn.

9. Click OK.

10. In the table of contents, right-click the layer name and open the attributes table. Notice that now only three columns are displayed. The other columns are still there but hidden. Just hiding the columns might work for some projects, but for many other projects you really may only want these three columns.

11. Close the attributes table.

12. Let's essentially make a copy of this file, and in so doing, capture only these three columns. Right-click the layer name in the table of contents, select Data, and then select Export Data.

13. Click the yellow file folder and navigate to where you would like to save what will be your new, slimmer file. Give it a new name, such as **MiniFile**.

14. Click Save.

15. Click OK and Yes to add the file to the map as a layer.

16. Open the attributes table to verify you have only the columns you exported (plus a couple of others that are standard).

7 Work with multiple attribute tables

Attribute tables are handled slightly differently in ArcGIS 10. You can open multiple attribute tables and toggle between them using tabs in the lower right corner of the table window.

1. Open both attribute tables (AgeJoined and MiniFile) if they aren't already. **4**

2. Click on each of the tabs in the lower left corner of the table's window and notice how you can switch between tables. The tabs are also movable, by dragging and dropping.

3. You can also dock the tables (new to ArcGIS 10) to create a split screen. Left-click and drag the MiniFile tab to the center of the tables window. A blue circle with four arrows will appear. Drop the tab on the right arrow. The screen should split with one table on each side. **5**

4. To undo the split screen, simply drag and drop the table back in the tab position.

4

	01127	Walker	14.7%
	01043	Cullman	14.6%
	01023	Choctaw	14.4%
	01031	Coffee	14.3%
	01095	Marshall	14.3%
	01037	Coosa	14.2%
	01133	Winston	14.2%

MiniFile AgeJoined

5

Butler	0.164307
Choctaw	0.144077
Fayette	0.161828
Jackson	0.133016
Winston	0.141891
Baldwin	0.154
Autauga	0.1
Barbour	0.134
Calhoun	0.140981
Clay	0.16683
Greene	0.150291
Colbert	0.15419

GOOD TO KNOW

You can also click the query icon in the table toolbar to write an attribute query. Attribute queries will be covered in chapter 15.

CHAPTER 8

Address mapping

Address mapping, also called geocoding, is like creating a pushpin map of addresses. Geocoding is a skill everyone is likely to need in their GIS work. You might use this skill for mapping things like diseases, crimes, client addresses, service addresses, or anything with a physical location—an address—that you would like to show on a map.

It's helpful to have a conceptual understanding of how geocoding works before attempting to do it. The most important thing to know is that geocoding is a two-step process: pregeocoding and geocoding. Pregeocoding involves setting up an address locator and should be thought of as building the foundation on which to geocode. When you set up an address locator, you are predetermining all the rules that will govern the geocoding session (such as spelling sensitivity). The second part of the process, geocoding, is actually placing the addresses on a map.

Exercise goal

Geocode social service agency addresses. ①

Exercise file locations

Chapter directions: Follow the exercise as it appears in this book

This exercise uses the following:

- An Excel spreadsheet containing social services in Bexar County, Texas. This file has been provided on this book's accompanying CD. The file is called SocialServiceAddresses.xls and is located in the chapter 8 folder.
- A street shapefile for Bexar County, Texas. The file is downloaded as a part of this exercise.

Note: You will use the geocoded_result.shp created in this exercise in the next chapter about categorical mapping. If you intend to do the next exercise, please note where you save this file.

CD: Use the CD included with this book

All files needed for this exercise are included on the book's CD. Files are organized by chapter.

Personal files: Use files you've gathered from other sources

- To complete this exercise, you will need an Excel spreadsheet containing addresses. Your spreadsheet must look like the one below. The most important factor is to have the addresses in one column called Address. You do not need to have ZIP Codes for your addresses, but you will end up with more accurate matches if you do.
- You will also need a street shapefile for the same geography your addresses cover.

ORG_NAME	ADDRESS	CITY	STATE	ZIP
Happy Days Homecare	4854 Shadydale Drive	San Antonio	TX	78228
Amity House	4855 West Commerce Street	San Antonio	TX	78237
Council 4 Independent Living	4905 Center Park Boulevard	San Antonio	TX	78218
Nereida Cardona	4906 Ali Avenue Apt. 1	San Antonio	TX	78229
Hills Child Care	4914 Frostwood Drive	San Antonio	TX	78220
Daniel P Diaz Ph.D.	4939 Devasales Road	San Antonio	TX	78249
Jeannie OSullivan RNC Lpc	4940 Broadway Street Ste. 230	San Antonio	TX	78209
Kidcare Drop-In Child Care	4951 Northwest Loop 410	San Antonio	TX	78229
Pilgrim Congregational School	500 Pilgrim Drive	San Antonio	TX	78213
Nccj	501 South Main Ste. 101	San Antonio	TX	78204
Angels Daycare	5010 Lancelot Drive	San Antonio	TX	78218
Adoptive Family Support Svcs	5018 San Pedro Avenue	San Antonio	TX	78212
St Johns Lutheran Church	502 East Nueva	San Antonio	TX	78205
Ultracare Independent Living	502 K Street	San Antonio	TX	78220

1 Go to the U.S. Census Bureau Web site and select files

1. Go to www.census.gov.

2. On the main Census site, to the right of the Geography link, select the TIGER link.

3. The first link on that page is the gateway to the most current census shapefiles. Select the 2009 TIGER/Line Shapefiles Main Page link.

4. On the next page, select Download Shapefiles on the left in the orange TIGER Navigation panel.

2 Select geography (Bexar County, Texas, All Lines file), download, and unzip

1. Click the Download 2009 TIGER/ Line Shapefiles now link.

2. Instead of a layer for the entire nation (as you would get if you select a link on the left), select a street network for Bexar County, Texas. Select Texas from the State- and County-based shapefiles on the right and click Submit. Do not click any check boxes on the left.

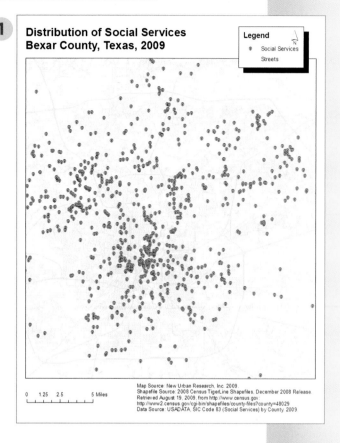

Distribution of Social Services Bexar County, Texas, 2009

Legend
• Social Services
Streets

0 1.25 2.5 5 Miles

Map Source: New Urban Research, Inc. 2009.
Shapefile Source: 2008 Census TigerLine Shapefiles, December 2008 Release.
Retrieved August 19, 2009. from http://www.census.gov.
http://www2.census.gov/cgi-bin/shapefiles/county-files?county=48029
Data Source: USADATA, SIC Code 83 (Social Services) by County. 2009.

3. From the list of Texas county files, select Bexar and click Submit.

4. Double-click the All Lines link. Do not use the check box option.

5. When asked whether you want to Open, Save, or Cancel the file, click Save and save the zipped file on your C drive in a folder called **Geocoding**.

6. Unzip the files. You can use any unzipping program you wish. Follow the instructions for that program to unzip the first file then the second file. If you do not know if you have an unzipping program, try double-clicking the zipped file to unzip it.

MORE ABOUT THE ALL LINES FILE

The All Lines file (which is named the edges file upon download) primarily contains all the streets in a county. However, the file includes several other things such as line features for water, railroads, and trolleys. This file includes anything that has a linear edge to it.

The MTFCC column contains a key to what each feature represents. Appendix F of the TIGER/Line Technical Documentation contains the key to the MTFCC.

Part 1: Pregeocoding (or setting up an address locator)

3 Open ArcGIS then ArcCatalog

1. Open ArcGIS.

2. Open ArcCatalog by selecting the Windows menu (at the top) and selecting Catalog. (You can also select the Catalog icon 🗺). ArcCatalog will look slightly different in older versions of the software. **2**

4 Define a geocoding style

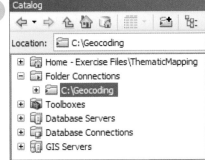

> ### GEOCODING WHEN YOU DON'T HAVE ZIP CODES
>
> If you do not have ZIP Codes as part of the address file, you can still geocode. Simply select U.S. Streets as the geocoding style.

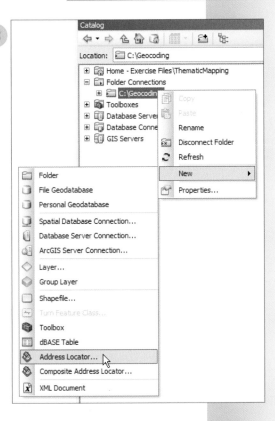

1. This part is a little confusing. In the ArcCatalog window, navigate to a folder where you would like to save what will be your newly created Address Locator (perhaps where you saved your street network). Once you navigate to that folder, right-click on the folder name, select New, then select Address Locator. **3**

2. In the first field Address Locator Style, click the Browse Folder icon 📁 and select US Address-Dual Ranges. A warning message will appear at the top in yellow, but we'll fix that in a few steps.

3. In the Reference Data field, click the browse folder icon and navigate to, then Add, the streets network file (**tl_2009_48029_edges.shp**) on your computer (or network), which you just downloaded from the Census. Here you'll see another red circle indicating an error.

4. Under the Alias Name column, change the following fields from None to their aliases, as indicated below. The first four should already be filled in, leaving you to fill in **FULLNAME** for the Street Name. In other versions of the software, you must fill in all the fields indicated.

   - House From Left to LFROMADD
   - House To Left to LTOADD
   - House From Right to RFROMADD
   - House To Right to RTOADD
   - Street Name to FULLNAME

5. The error messages should disappear once all the aliases are filled.

6. Scroll down to the last field, Output Address Locator, click the yellow browse folder to navigate to the Geocoding folder on your C drive. Give it the name **This is my Address Locator**, so it will be easy to spot when we next need it. Your address locator should look like the screenshot here, although your file paths for reference data and output address locator fields will likely be different. ④

7. Click Save and OK.

8. The geocoding processor will take a few seconds to process the address locator before a confirmation message appears. If you expand the folder contents, you should be able to see a little red icon. This is your new Address Locator.

9. Close ArcCatalog.

Part 2: Geocoding

5 Open files

1. Use the Add Data button ✛ ▾ to add the street network shapefile (**tl_2009_48029_edges.shp**).
2. Use the Add Data button to add the Excel file (**SocialServiceAddresses.xls**). You have to double-click the Excel file and add the worksheet **SocialService Addresses$**. Both should appear in the table of contents.
3. Right-click the Excel file and select Open, just to review the addresses that you are about to geocode. You should see 1,144 social service addresses that will be plotted out against the street network in the lower right corner.
4. Close the table.

6 Geocode addresses

1. In the table of contents, right-click the addresses file name. Select Geocode Addresses.
2. Select the Address Locator you just created and saved, called **This is my Address Locator**. (If you do not see the address locator displayed, you will need to click the Add button and navigate to the address locator.) **5**
3. Click OK.
4. In the next dialog box, leave the default settings as they are with the exception of where to save the output file. **6** Save it in a convenient place on the C drive. The default name of the new file will be **Geocoding_Result.shp**.
5. Click the Geocoding Options button and change the Minimum Catch Score to 75. Click OK twice.
6. The geocoder will geocode your addresses. The Geocoding Addresses dialog box provides statistics about what happened in the geocoding session. The most important thing to be aware of is the unmatched rate. A general guideline is that the unmatched rate should be 5

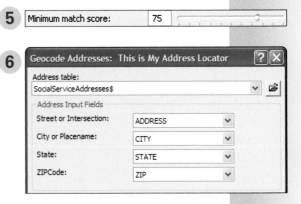

5 Minimum match score: 75

6 Geocode Addresses: This is My Address Locator

Address table:
SocialServiceAddresses$

Address Input Fields

Street or Intersection: ADDRESS

City or Placename: CITY

State: STATE

ZIPCode: ZIP

PROBLEM ALERT

If you get an error message stating "There was an error trying to process this table," which prevents you from geocoding, the problem is with one of your column types.

To fix this, in the table of contents, right-click the addresses file and select Properties. Select the Fields tab and change any numeric fields to an integer field, and try regeocoding. This error occurs frequently when currency fields are included in a table you are trying to geocode.

Problems also occur if the street network is unprojected. Your street network must have a datum (such as NAD83) associated with it. The street networks from the census come with a datum of NAD83.

percent or less. Since we're at 14 percent we need to manually rematch some of the unmatched ones. **7**

7. Click Rematch. (If you accidentally clicked Close, right-click the geocoded file in the table of contents, select Data, and Review/Rematch Addresses).

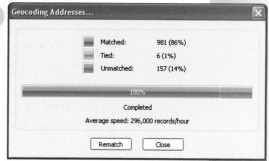

7 Manually geocode unmatched addresses

1. After you click Rematch, an interactive screen will appear. This is where we'll attempt to manually geocode addresses that did not match during the automatic geocoding. First, make this screen larger so you can see all the elements (drag the lower right corner of the box to resize).

2. Select just the unmatched addresses from the menu at the top. **8**

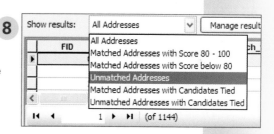

3. At this point, you might be unclear as to what to do. The first record you see, the one highlighted in light blue, is the first unmatched address. But you won't be able to see the address unless you scroll to the far right (or shorten the column Match_addr, as it's very wide). Scroll over so you can see the actual unmatched address (10010 Broadwa Street Apt. 805).

8 Fix addresses

1. To widen the Address section so you can see the address clearly, use your mouse to drag the light gray line to the right of this section. **9**

2. In the field under Address, type over the given address. In this example, the geocoder is confused because the street name is misspelled. In the Street or Intersection box add a "y" to the end of Broadwa.

3. Click Search at the bottom of the screen.

4. The candidates listed will refresh with new matches. The first has the highest score and is usually the best option. In this case, it is the correct address.

There is no magic number for what the lowest acceptable score should be, and it depends on the accuracy requirements of your analysis. First note your address, then look at the address ranges and find where the closest match is. It is a subjective decision about whether to match it or not.

5. To put a dot on the map where the address is located, select the first candidate and click the Match button. **10**

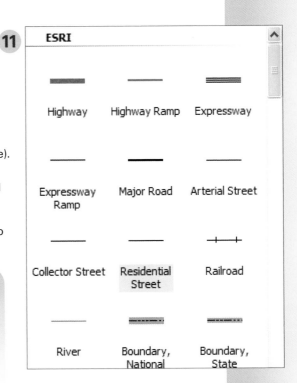

6. To continue matching other addresses, simply select the next unmatched address and repeat the process. Try this with a few others just to get the hang of it.

7. Once you are finished, click Close. Take a look at your map.

9 Improve the look and learn about symbols (optional)

ArcGIS assigns random colors to shape-files. Let's change the color of the streets and the shape and color of the dots.

1. In the table of contents, right-click the street shapefile name, and select Properties and the Symbology tab.

2. Click the color patch (which is a colored line).

3. Notice now that all sorts of street templates are given. Select the residential one (this makes the lines very skinny) but change the color of the line from black to the lightest gray. This will make it easier to read. Click OK twice. **11**

GLOBAL SEARCH AND REPLACE

Let's say you have the same street name misspelled throughout your list of addresses. Instead of having to correct each one of these manually, you can make the attributes table editable (see chapter 7, step 3) and use the Find/Replace tool to find and replace misspelled text.

4. Now do the same thing for the geocoded file.

5. Select a shape, such as a triangle or square, and notice how your dots change. You can also experiment with the Style References button. Click Style References and select Crime Analysis by checking the check box next to it. You see many new and interesting symbols. Here you may want to experiment and find symbols related to your industry.

GOOD TO KNOW

You may have noticed several unmatched addresses after your geocoding session. It would take some time to fix all of these addresses manually. Instead, you may choose to export the attributes table of the geocoded shapefile and edit in Excel. To do this, sort the Status column descending with the "U" values at the top. These are the unmatched addresses (M means matched). Highlight the records you would like to export, select the Table Options button in the upper left corner, then select Export. Export all records as a dBase table (.dbf). You can fix these in Excel and regeocode them.

CHAPTER 9

Creating a categorical map

Categorical mapping is similar to thematic mapping in that color shading is used to indicate values. However, with categorical mapping the values represent categories instead of numbers.

This technique can be used with polygons, for example, to map land use zoning categories (residential, commercial, and industrial). It can also be used with point data, such as a geocoded file, to map such things as crime (burglaries, assaults). Categorical mapping also works with line data to map different types of streets (residential, major arterial, and highway).

Exercise goals

Categorically map point data (social service agencies by type) and polygon data (counties by name). ⬤1

Exercise file locations

Chapter directions: Follow the exercise as it appears in this book

The first part of this exercise uses the social services geocoded file created in the chapter 8 exercise (geocoding_result.shp).

The second part of this exercise used the **AgeJoined.shp** file used in chapters 5, 6, and 7.

CD: Use the CD included with this book

All files needed for this exercise are included on the book's CD. Files are organized by chapter.

Personal files: Use files you've gathered from other sources

To complete this exercise, you will need the following:

- A point shapefile (such as a geocoded file) that contains some sort of categorical data such as types of services, diseases, or crime. This is for the first part of this exercise.
- The second part requires a polygon file. You must have a shapefile that contains a column with some sort of categorical data such as names of places or type of place (not a continuum of data as was used in thematic mapping).

1 Add a shapefile and review the attributes

1. Open ArcGIS.
2. Click the Add Data ➕ ▾ icon.
3. Either open Folder Connections or select the Connect to Folder icon and navigate to **geocoded_results.shp** that you created in the last chapter.
4. In the table of contents, right-click the layer name and select Open Attributes Table.
5. Scroll to the far right and review the SICDESC column. This column represents the type of service each organization provides. SIC stands for Standard Industry Classification (which is the type of business performed) and DESC stands for Description. If you look at the column, you'll see several types of services provided. This is the type of information (names of things) that lends itself to categorical mapping.
6. Close the attributes table.

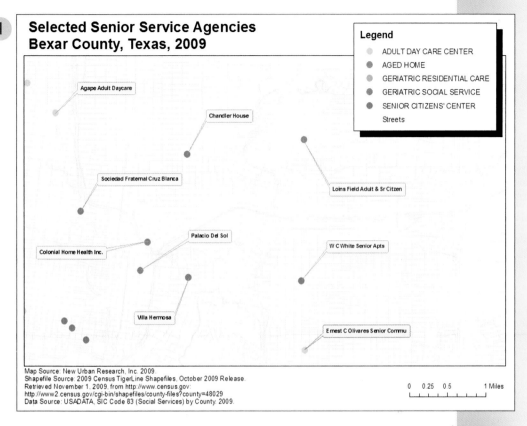

**Selected Senior Service Agencies
Bexar County, Texas, 2009**

Legend
- ADULT DAY CARE CENTER
- AGED HOME
- GERIATRIC RESIDENTIAL CARE
- GERIATRIC SOCIAL SERVICE
- SENIOR CITIZENS' CENTER
  Streets

Agape Adult Daycare

Chandler House

Sociedad Fraternal Cruz Blanca

Loins Field Adult & Sr Citzen

Palacio Del Sol

Colonial Home Health Inc.

W C White Senior Apts

Villa Hermosa

Ernest C Olivares Senior Commu

Map Source: New Urban Research, Inc. 2009.
Shapefile Source: 2009 Census TigerLine Shapefiles, October 2009 Release.
Retrieved November 1, 2009, from http://www.census.gov:
http://www2.census.gov/cgi-bin/shapefiles/county-files?county=48029
Data Source: USADATA, SIC Code 83 (Social Services) by County. 2009.

0   0.25   0.5   1 Miles

2 Create a categorical map

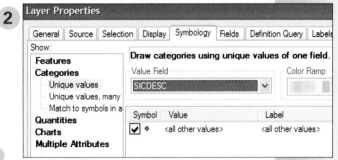

1. In the table of contents, right-click the shapefile name, select Properties and the Symbology tab.

2. From the list on the left, select Categories and Unique values.

3. In the Value field select the column of data you want to display on the map, SICDESC. 2

Layer Properties

General | Source | Selection | Display | Symbology | Fields | Definition Query | Labels

Show:
- **Features**
- **Categories**
  - Unique values
  - Unique values, many
  - Match to symbols in a
- **Quantities**
- **Charts**
- **Multiple Attributes**

Draw categories using unique values of one field.

Value Field   Color Ramp
SICDESC

Symbol	Value	Label
☑ ◇	<all other values>	<all other values>

4. The symbol can be changed to anything you like. For this exercise, a smooth circle is preferable to the default dot.

   To change the symbol, double-click the dot (next to the check box). For the symbol selector, choose Circle1 (you may need scroll down depending on what symbol palettes were selected in the last chapter) and change the size from 18.00 to **9**.

5. Click OK.

6. Click the Add All Values button and select a color ramp (upper right corner).

7. Click OK to see the colors displaying the data on your map. **3**

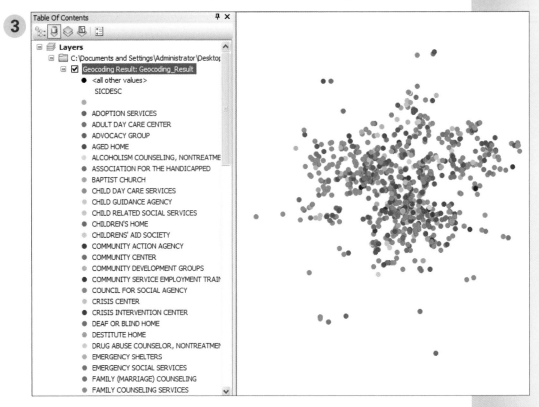

The map will look like someone threw confetti all over it. With this many categories and colors it's hard to read the map clearly. Let's select only a few types of services to display.

3 Display only certain types of agencies

1. In the table of contents, right-click the shapefile name, and select Properties and the Symbology tab.

2. Select Remove All to get rid of all the selections.

3. Click Add Values.

4. Click Complete List. When you have a lot of things on your list, sometimes ArcGIS shortens the list. If the Complete List button is available, you should go ahead and select it to ensure you are seeing all of list items.

Select the following five types of service: Adult Day Care, Aged Home, Geriatric Residential Care, Geriatric Social Service, and Senior Citizens' Centers. (You can hold down the Ctrl key to select more than one item at one time, or just select items individually and keep adding.) Click OK.

5. Make sure the check box next to All Other Values is not checked (as we only want to display the five types we've selected). Click OK.

6. You can change the colors of any of the dots by double-clicking the dot and making the change.

4 Group several categories into one category

You can also assign multiple categories to one all encompassing category like Senior Services.

1. In the table of contents, right-click the shapefile name, and select Properties and the Symbology tab.

2. Highlight all five categories by holding down the Ctrl key as you click each category name. **4**

3. Right-click within the dark blue highlighted area and select Group Values.

4. All five categories should now be grouped into one category (the name at this point doesn't matter).

5. Click OK.

6. In the table of contents, left-click once on the grouped layer to activate the text. Rename this grouping **All Senior Services**. Click OK.

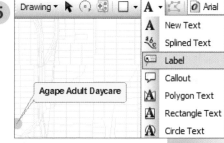

7. (Optional) To put labels on some of these dots, you can use the labeling tool, accessible via the Draw toolbar under Customize, Toolbars. After the tool is activated, expand the capital A. Click the Label tool to activate it, and then click a dot on the map to label it. **5**

8. (Optional) The map that is illustrated on page one of this chapter has callout boxes around the labels. To apply these, with the default pointer ▸, select all the labels (hold down the Ctrl key to select more than one). After they are selected, right-click any one of the labels and select Properties. Select Change Symbol then Edit Symbol. Select the Advanced Text tab and check the check box next to Text Background. Click the Properties button underneath it. Select the second balloon callout, and click OK four times. You may need to reposition the labels to make the anchor visible.

Categorical mapping with names

In the next steps, we'll assign a unique color to individual counties.

5 Close ArcGIS and reopen

Close ArcGIS completely and reopen (we need a fresh workspace for the next part of this exercise).

Add the AgeJoined.shp shapefile using the Add Data button.

6 Create a categorical map from polygons

1. In the table of contents, right-click the shapefile name, and select Properties and the Symbology tab.
2. Select Categories and Unique Values.
3. In the Value field, select County as the column of data you would like to see displayed on the map.
4. Click Add All Values and select a color ramp.
5. Click OK to see the colors displaying the data on your map.

7 Display specific counties

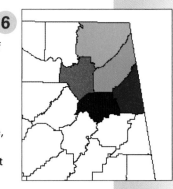

Let's say you want to color five counties in the northeast part of the state.

1. In the table of contents, right-click the shapefile name, and select Properties and the Symbology tab.
2. Select Remove All to get rid of all the selected counties.
3. Click Add Values and select the following five counties: Cherokee, DeKalb, Etowah, Jackson, and Marshall. You can hold down the Ctrl key to select more than one county at one time, or just select them individually and keep adding. Click OK.
4. Double-click the color patch next to each county and assign your own color (by using the Fill Color button).
5. Make the fill color for "all other values" solid white.
6. Click OK to see the colors displaying the data on your map. 6

TRANSPARENCY

Sometimes with shapefiles it may be helpful to make the color shading semitransparent—for example, if you wanted to overlay other shapefiles on top of this one, but you want both layers to be colored. To do this, right-click the layer name in the table of contents and select Properties. Select the Display tab and choose 50 percent for Transparent.

CHAPTER 10

GPS point mapping

In addition to mapping a table of addresses (see chapter 8), you can also map a table of latitude and longitude (x,y) coordinates collected in the field using a Global Positioning System (GPS) unit. Since each GPS unit is different, it's difficult to explain how to get the GPS points out of your handheld device and into Excel. At the end of this chapter, some general guidelines are given that may assist you.

Latitude and longitude points are frequently used in GIS, such as when a particular place doesn't have an address, but you still need to display it as a dot on the map. Industries such as transportation and utilities often need to click points out in the field and associate data with those points. People doing varied work on the environment do, too.

Exercise goal

Place latitude and longitude points on a map.

Exercise file locations

Chapter directions: Follow the exercise as it appears in this book

This exercise uses the following:

- An Excel spreadsheet containing latitude and longitude values for social services in Bexar County, Texas. This file is provided on this book's accompanying CD in the chapter 10 folder. It is called **XYData.xls**.
- A street shapefile for Bexar County, Texas. This is the same file used in chapter 8. If you didn't do that exercise, you can download it from the Census site using the instructions in chapter 8.

CD: Use the CD included with this book

All files needed for this exercise are included on the book's CD. Files are organized by chapter.

Personal files: Use files you've gathered from other sources

To complete this exercise, you will need the following:

- An Excel spreadsheet containing latitude and longitude values, like those you might download from a handheld GPS. Your spreadsheet should look like the one below. The most important thing is to have an X column (longitude) and Y column (latitude).
- A correctly projected shapefile of the general area where your points will be located (such as the city or county). It doesn't have to be exact in terms of the location. We'll use streets in this exercise.

	F25	▾	fx	
	A	B	C	D
1	X	Y	ORG_NAME	SICDESC
2	-98.7815079439000000	29.4154714690394000	Bright Horizons	CHILD DAY CARE SERVICES
3	-98.7303817441000000	29.4206082558429000	Christian Hands Child Care	CHILD DAY CARE SERVICES
4	-98.7056629219000000	29.5419655273119000	Rainbow Hse For Women Children	EMERGENCY SHELTERS
5	-98.6991906328000000	29.4470557396205000	Saint Joseph Foster Homes	RESIDENTIAL CARE FOR CHILDREN
6	-98.6969255182000000	29.4098258714620000	Freda M Lee	CHILD DAY CARE SERVICES
7	-98.6921489629000000	29.4298936476014000	Happy Tots Home Care	CHILD DAY CARE SERVICES
8	-98.6899312602000000	29.5751458661927000	Helotes Y School Age	CHILD DAY CARE SERVICES
9	-98.6895279650000000	29.4283065918299000	After School Kare Ed Cody	GROUP DAY CARE CENTER
10	-98.6890343094000000	29.5668876883472000	Helotes Counseling Center	GENERAL COUNSELING SERVICES
11	-98.6869799105000000	29.5088181177161000	Rosies Child Care Center	CHILD DAY CARE SERVICES
12	-98.6804988165000000	29.4323077615336000	Alloy Wheel Repair Special	HELPING HAND SERVICE (BIG BROTHER, ETC.)
13	-98.6804396202000000	29.3129819752082000	Cils Day Care	CHILD DAY CARE SERVICES
14	-98.6791796419000000	29.4927075129563000	Childtime Learning Center	CHILD DAY CARE SERVICES
15	-98.6782709162000000	29.5247223483571000	Tabatha L Bensel	CHILD DAY CARE SERVICES
16	-98.6777633474000000	29.4131049337576000	Diana Roberts Day Care Ctr	GROUP DAY CARE CENTER
17	-98.6772952838000000	29.4344006133606000	First Aid Remodeling	GERIATRIC SOCIAL SERVICE

1 Add a shapefile of a general area where your dots should be placed

1. Open ArcGIS.
2. Click the Add Data icon ✛ ▾ .
3. Either open Folder Connections or select
   the Connect to Folder icon and navigate to
   tl_2009_48029_edges.shp, which was used in
   chapter 8.

GOOD TO KNOW

You don't always have to use a street network.
The file you use should be the location of
where the x,y coordinates will be located. For
example, if you collected points in your city,
then a city shapefile would work well.

Courtesy of Trimble

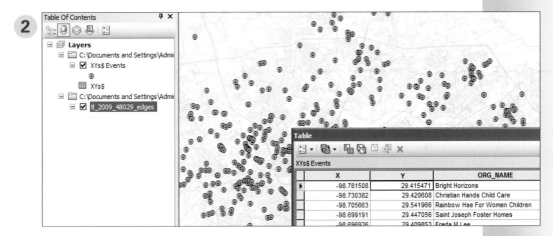

2 Open the Excel file with longitude and latitude points and review

1. Next you want to add the Excel file to the ArcGIS window. Click the Add Data button and navigate to the file and *double*-click it, then select the worksheet (denoted with a $ sign). The file is called **XYFile.xls** and the worksheet is **XYs$**.

2. In the table of contents, right-click the file and select Open. Have a quick look at the data table to confirm that you have one X (longitude) column and one Y (latitude) column (as well as a couple of other columns).

3. Close the data table.

3 Add X,Y data and define the X field and the Y field

1. In the table of contents, right-click **XYs$** and select Display X,Y Data.

2. The X and Y fields automatically fill in based on the column headers of X and Y. There is no Z field.

3. The coordinate system must be defined. Click Edit then click Select. Browse for and select the appropriate datum that pertains to your original GPS source file. In this example, navigate to the Geographic Coordinate System folder, North American, and then double-click NAD 1983.prj (this is a common datum to use). Click Add. For more information on projections, see chapter 3. Click OK twice. ③

4. When you get a message stating "Table does not have Object-ID field," click OK. It will still open with no problems.

4 Create a new shapefile of XY Events

1. Right now the X,Y layer is not saved on the C drive, it's only a temporary file. To create a new, permanent shapefile of these X,Y points, in the table of contents, right-click the **XYs$ Events** layer.

2. Select Data, select Export Data, and save your new shapefile (name it **XYPoints**).

GETTING GPS POINTS OUT OF YOUR GPS UNIT AND INTO EXCEL

Every GPS unit is different, so it's difficult to give specific advice. The general idea is that you will need to export the points out of your GPS unit (perhaps as a text, Comma Separated Value, or database file), open them in Excel, do some minor cleanup work, and save.

1. Click points in your GPS, sometimes called waypoints or points of interest (POIs).

2. Connect your GPS to your computer via the USB cord.

3. Your GPS likely came with software that must be installed on your computer ahead of time. If it has been installed, you can navigate to the folder where your clicked points are stored in your GPS.

4. Often each GPS unit uses a proprietary file format, so don't be surprised if you don't recognize the file extension of your points file.

5. You should be able to open the file in Notepad to make sure it's the correct file. You might try double-clicking the file and when prompted for which software to use to open it, select Notepad. Save as a text file (.txt). Close file.

6. Open Excel. Open your text file in Excel. This process will vary depending on what version of Excel you are using. You may need to import the file into Excel, or you may simply be able to open the file.

7. Once your data is imported into Excel, identify your X (longitude or easting) and Y (latitude or northing) coordinate fields, and any corresponding attribute fields. If needed, add or modify the header row to give the columns easy-to-understand titles. Longitude should be titled X and latitude should be titled Y.

8. Save as an Excel file and name the worksheet. Now the file is ready to open in ArcGIS.

CHAPTER 11

Editing boundaries

Occasionally you may need to change the physical boundary of an existing polygon. For example, if your agency uses a target area boundary to deliver services and would like to extend the service area, editing the boundary would become necessary. Target area boundaries might include things like school districts, voting wards, or neighborhoods.

In this exercise, we will change the boundary of a state polygon (we use Oregon but you can use your own state). Of course in real life, it's highly unlikely that you would need to change a state boundary. In this exercise, though, we chose a simple and easily recognizable boundary to make it as easy as possible to learn how to edit boundaries. **1**

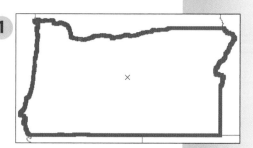

Exercise goal

Perform various editing tasks, including changing a boundary outline, merging polygons, creating shapefiles out of selected polygons, appending shapefiles to each other, and clipping shapefiles.

Exercise file locations

Chapter directions: Follow the exercise as it appears in this book

This exercise uses the following:

- A U.S. states shapefile.
- A U.S. counties shapefile.

These files are downloaded as a part of this exercise.

CD: Use the CD included with this book

All files needed for this exercise are included on the book's CD. Files are organized by chapter.

Personal files: Use files you've gathered from other sources

To complete this exercise, you will need at least two shapefiles that you don't mind editing.

1 Go to the U.S. Census Bureau Web site and select files

1. Go to www.census.gov.

2. On the main census site, to the right of the Geography link, select the TIGER link. A link on that page is the gateway to the most current Census shapefiles.

3. Select the 2009 TIGER/Line Shapefiles Main Page link.

4. On the next page, select the Download Shapefiles link on the left in the orange TIGER Navigation panel.

2 Make geographic selections (States and Counties for U.S.), download, and unzip

1. Under Nation-Based Shapefiles on the left, click the State and Equivalent (Current) link, which will give you one shapefile with all the U.S. states.

2. Click Save when prompted and save the zipped file on your computer's hard drive.

3. Under Nation-Based Shapefiles on the left, click the County and Equivalent (Current) link, which will give you one shapefile with all counties in the United States. This file is large and may take a while to download.

4. Click Save when prompted and save the zipped file on your computer's hard drive.

5. Unzip the files. You can use any unzipping program you wish. Follow the instructions for that program to unzip the first file then the second file. Each file should unzip in its own folder. If you do not know if you have an unzipping program, you might try double-clicking on the zipped file to unzip it.

One of the most common editing tasks is to change the shape of a boundary. The next two steps show you how to change the physical outline of a polygon.

3 Open a shapefile with polygons and turn on the Editor toolbar

1. Open ArcGIS.

2. Click the Add Data icon ✛ ▾ .

   Optional: Right-click inside the blank data frame and select Properties. Change the data frame's projection to "North America Albers Equal Area." Refer to chapter 3 for a reminder about projections. ⑦

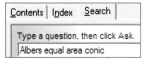

3. Either open Folder Connections or select the Connect to Folder icon and navigate to **tl_2009_us_state**, which you just downloaded and unzipped.

4. Use the Magnifying tool to zoom in closer to Oregon (or whichever state you want) so you can really see the outline well.

5. Click the Editor Toolbar icon 🖉 . In ArcGIS 10 this button should already be visible as it's a part of the default ArcGIS interface. In older versions of the software, you must go to View, select Toolbars and Editor. The icon will look the same.

6. Once you click the Editing icon, a new Editor toolbar should be visible. Dock the toolbar if it is not already docked by dragging it to the top of the window where the other toolbars are located.

4 Edit the state outline

1. Click the Editor button and select Start Editing. This makes not only the polygon boundaries editable, but also the attributes table.

2. A little arrow is now activated and serves as the pointer. With this arrow, *double-click* the polygon you would like to change, in this case Oregon, and notice how several little dots (ArcGIS calls them vertices) appear in green. ② Zoom in very close so you can really see them. By moving these dots, you can reshape the boundary of the state. Use the Magnifying tool to zoom in super close to be able to see the dots clearly. ③

You may need to re-activate the Edit tool ▶ after using the Magnifying tool.

3. Click any one of the dots and drag it to a different position to begin reshaping. **4** Try this with a few more vertices. When you are finished, click anywhere outside the state boundary to complete the edits.

4. To stop editing and save, click Editor and select Stop Editing, and then Save Edits when prompted.

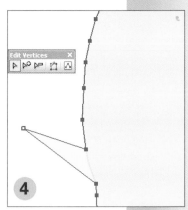

5. One last thing to know is that you can move an entire polygon at once. Click Editor again and click any polygon once. Notice the state becomes highlighted in light blue. This means the entire polygon is selected and you can actually move that polygon. Try moving it. Select Stop Editing and do not save changes.

Another common editing task involves merging multiple polygons into one large polygon *with the same outer boundary* as the first set of polygons. With merging, any data within the attributes table will be summed to one record. Essentially you're making one polygon out of multiple polygons.

5 Merge polygons ?

1. Begin by making the shapefile editable. Click Editor and select Start Editing.

2. Hold down the Shift key and with the Edit tool, select multiple contiguous polygons such as three different states (in this example Oregon, Washington, and Idaho). You can tell they are selected when all appear with bright blue shading. **5**

3. Click Editor and select Merge.

4. Because the data for those states will be summed and dumped into one of the three states, it is necessary to select one state for all the states' data. When you get the dialog box that says Choose the feature within which other features will be merged, select one of the states. Oregon was selected for this example. Click OK.

5. Notice that all three states have merged into one.

6. Open the attributes table. Notice that now there is no Idaho and Washington. The data for these states has been summed for all polygons and is now warehoused in one line item (Oregon).

Union differs from merge in that the data is not summed to one record. Instead, each polygon keeps its own record even though the visible boundary between the polygons disappears. *Visually* a union looks like one big new polygon (similar to merge), but within the attributes table, a record is maintained for each polygon.

6 Union polygons

1. Hold down the Shift key and with the Edit tool ▶, select another set of three multiple contiguous polygons. You can tell they are selected when all appear with bright blue shading. In this example, Texas, New Mexico, and Oklahoma are selected.

2. Click Editor and select Union. When the dialog box displays that says Choose a template to create feature(s) with, click OK.

3. Notice the three states now appear as one. However, they are still three distinct states. Click anywhere outside the states to get rid of the selection (the blue highlight). Now click more or less in the area where Texas should be (or any one of the states you may have selected) and notice how the original outline of that one state appears selected. Visually it appears to be one boundary, but they are still three separate line items in the attribute table.

4. Open the attributes table to confirm that you still have a line item for Texas, New Mexico, and Oklahoma. ⑦

The ability to select individual items in your maps and create a brand new file *with just those specific items* can be very useful. In this example, you will export California, Nevada, Utah, and Arizona to their own shapefile to create a focus area or target area.

7 Use the Select Features tool and create a new shapefile from selected geographies

1. On the toolbar at the top of ArcGIS, activate the Selections Tool button. 🔖 ▾

2. Hold down the Shift key and click once on each state to select it (California, Nevada, Utah, and Arizona). They will be highlighted in bright blue.

3. In the table of contents, right-click the shapefile **tl_2009_U.S._state**.

4. Select Data and Export Data. Type the file name **Southwest** and where to save it by clicking the browse folder icon. Click Save and OK.

5. Select Yes to add the data as a layer. This should add a new layer to ArcGIS.

6. To get rid of the blue outline, select the Selection menu at the top and Clear Selected Features. Notice if you check the check box next to each layer in the table of contents you can see just the Southwest shapefile.

Another useful technique is to take multiple shapefiles and combine them into one shapefile. The Append tool is most helpful here.

8 Create other shapefiles to append

1. We need other files to work with to show how to append. Right now you have one new shapefile (Southwest). Let's create two more.

2. Review the toolbars at the top of the window, select the Selections Tool icon.

3. Click any other state once—in this example, Montana. It will be highlighted in bright blue.

4. In the table of contents, right-click the shapefile **tl_2009_U.S._state**.

5. Select Data then Export Data. Specify the file name (Montana) and where to save it by clicking the browse folder icon. Click Save, then OK.

6. Select Yes to add the data as a layer. This should add your new layer to ArcGIS.

7. We need one more file. Click any other state once. (In this example, we selected Wyoming.) It will be highlighted in bright blue.

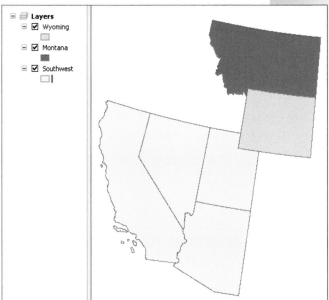

8. In the table of contents, right-click the shapefile **tl_2009_U.S._state**.

9. Select Data then Export Data. Specify the file name (Wyoming) and where to save it by clicking the browse folder icon. Click Save and OK.

10. Select Yes to add the data as a layer. This should add your new layer to ArcGIS.

11. To remove the state layer, in the table of contents right-click and select Remove. All you should have left open are the three shapefiles: Southwest, Montana, and Wyoming. 6

9 Create a new shapefile that will ultimately become the new layer containing all files

Now we're ready to create a new shapefile that will contain all three of these shapefiles. The first step is to make a new shapefile that will eventually hold all three files. Think of this as a starter file.

1. In the table of contents, right-click the Southwest shapefile layer name, select Data and Export Data.

2. Use the browse button to navigate to where you would like to save this file. Give this file the new name **AllThreeTogether.shp**.

3. When prompted, add it to the table of contents.

10 Activate the Append tool

1. Click the ArcToolbox icon 🖼 to open ArcToolbox.

2. Expand the Data Management Tool menu, then expand the General menu.

3. Double-click Append.

11 Append

1. In the first field Input Datasets, select the Wyoming and Montana shapefiles. Do not include the **AllThreeTogether.shp** or the **Southwest.shp**. (We created the **AllThreeTogether** file from the Southwest file, so it already has the Southwest geographies in it. If you do include it, duplicate records from that file will append. You are adding the Montana and Wyoming files to the **AllThreeTogether. shp**, which, as it stands, is composed of the Southwest polygons.)

2. In the Target Data set drop-down menu, select the **AllThreeTogether.shp** file. This is where they are going to be appended.

3. In the Schema Type, leave as the default TEST.

   NOTE: In the future, if the attributes table for each of your geographies is not identical, select NO_TEST.

4. Click OK. It may take a minute to complete.

5. Open the attributes table of **AllThreeTogether.shp** and review the results. Now you have one shapefile, created from multiple other files.

   Another tool you may find useful is Clip. You can clip one boundary by using the outline of another boundary. For example, let's say you have a shapefile of all the counties in the United States and you also have a boundary of Montana. Clipping allows you to select only those counties within the boundary of Montana. So you can create a new shapefile of just those counties in Montana. Let's try it.

12 Add data to be clipped

1. Close ArcGIS and reopen with a blank map. No need to save any changes. We need a clean, fresh mapping session.

2. Add the **Montana.shp** file and the U.S. counties shapefile (**tl_2009_U.S._county.shp**) to the ArcGIS window, using the Add Data button.

3. Zoom in so you can see Montana fairly well. You may have to move Montana in to the first position in the table of contents. **7**

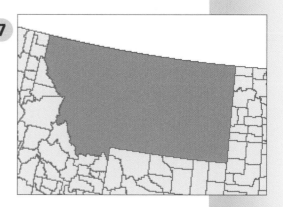

13 Clip

1. Under the Geoprocessing menu, select Clip.

2. For Input Features, select the file *to be clipped* (**tl_2009_U.S._county.shp**).

3. For Clip Features, select the file that will serve as the clipped boundary, in other words, the file that will be used to clip the first file (**Montana.shp**).

4. In the Output Feature Class drop-down box, navigate to where you would like to save what will be a newly created shapefile and give it a name such as **CountiesWithinMontana.shp**. **8**

5. Click OK. The operation may take a few minutes.

6. The new file will automatically be added to ArcGIS. Turn off all other shapefiles so you can clearly see the new **CountiesWithinMontana.shp** file. **9**

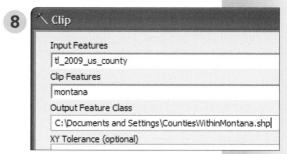

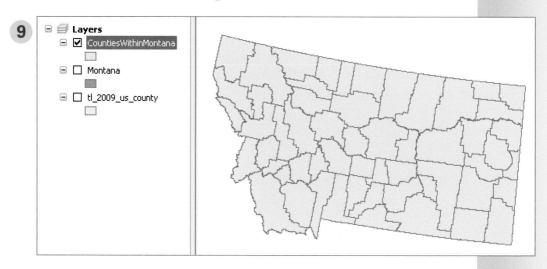

CHAPTER 12

Joining boundaries

In chapter 5 you joined a data table to a shapefile. In this chapter, you will perform a spatial join, which involves combining two shapefiles (and their data tables) into one. For example, what if you wanted to figure out which census tracts are in which neighborhoods in your city? One way to accomplish this is by clicking each tract on a map and writing down the neighborhood it's in. But this would take much too long!

Instead, you can perform a quick spatial join to combine tract and neighborhood shapefiles. The end result will be an attributes table that includes both tract and neighborhood data. This is an extremely powerful tool that is useful in many situations.

Exercise goal

Perform a spatial join between Alabama cities and counties shapefiles and create an attributes table that includes the county where each city is located. **1**

Exercise file locations

Chapter directions: Follow the exercise as it appears in this book
This exercise uses two shapefiles from chapter 1:
- Alabama counties (**tl_2009_01_county.shp**)
- Alabama cities (**tl_2009_01_place.shp**)

CD: Use the CD included with this book
All files needed for this exercise are included on the book's CD. Files are organized by chapter.

Personal files: Use files you've gathered from other sources
To complete this exercise, you will need any two shapefiles that you would like to join.

---

1 Add shapefiles

1. Open ArcGIS.
2. Click the Add Data icon ✛ ▾ .
3. Either open Folder Connections or select the Connect to Folder icon and navigate to the place shapefile (**tl_2009_01_place.shp**) and a county shapefile (**tl_2009_01_county.shp**) for Alabama.
4. You may need to drag and drop layers to different positions so you can see them, or you can turn each layer off and on.

2 Activate the Spatial Join tool

1. Click the ArcToolbox icon 🗔 to open ArcToolbox.
2. Expand the Analysis Tools menu, then expand the Overlay menu.
3. Double-click Spatial Join.

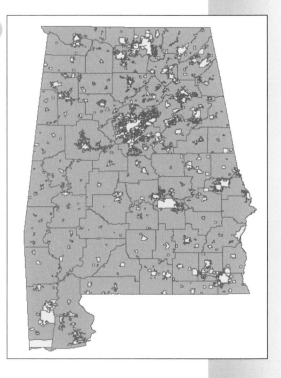
1

3 Fill in spatial join options

1. For Target Features, select the shapefile to which you want the data appended. This will be the places shapefile **tl_2009_01_place.shp**.

2. For Join Features, select the file that you would like to append. This will be the county shapefile **tl_2009_01_county.shp**.

3. For Output Feature Class, navigate to where you would like to save the newly created file on your C drive, and give it a name such as **SpatialJoin**.

4. In the Field Map section, you have the option of excluding certain fields. Just leave these fields as they are.

5. For the Match Option, select Intersects. There are other options, but the most commonly used is Intersect. Click OK. The operation may take a minute. The new SpatialJoin shapefile should have been added to the table of contents. **2**

Spatial Join

Target Features
C:\Documents and Settings\Administrator\Desktop\Exercise Files\tl_2009_01_place.shp

Join Features
C:\Documents and Settings\Administrator\Desktop\Exercise Files\tl_2009_01_county.shp

Output Feature Class
C:\Documents and Settings\Administrator\Desktop\Exercise Files\SpatialJoin.shp

Join Operation (optional)
JOIN_ONE_TO_ONE

☑ Keep All Target Features (optional)

Field Map of Join Features (optional)
- STATEFP (Text)
- PLACEFP (Text)
- PLACENS (Text)
- PLCIDFP (Text)
- NAME (Text)
- NAMELSAD (Text)
- LSAD (Text)
- CLASSFP (Text)
- CPI (Text)
- PCICBSA (Text)
- PCINECTA (Text)
- MTFCC (Text)
- FUNCSTAT (Text)
- ALAND (Double)
- AWATER (Double)

Match Option (optional)
INTERSECTS

4 Open attributes table and view results

1. In the table of contents, right-click the new shapefile name and select Open Attribute Table.

2. Notice a column was added that indicates the county for each city (column Name_1).

CHAPTER 13

Aerial photography

Shapefiles are called vector data while aerial photographs are classified as raster data. The key difference is raster maps are drawn with pixels and have no associated latitude and longitudinal data. Raster maps are not considered "smart maps" because you cannot click a part of the map and get information about that particular feature. On the other hand, raster maps are great for displaying a photographic snapshot of what is happening on the ground in any particular area.

Aerial photography can be particularly useful when combined with shapefiles. For example, utility companies may want to view a picture of a five-block area, and then overlay (in shapefile format) utility poles (points perhaps collected from a GPS) and associated information about those poles such as electricity generated or malfunctioning power lines. This type of information changes constantly and can easily be updated in a shapefile, but not in an aerial photograph. Therefore, learning to work with both types of files, especially in combination, is useful.

In this exercise we're focusing on aerial photographs; in the next exercise we'll work with a scanned paper map image. The two processes are similar but not the same.

Exercise goal

Import an aerial photograph into ArcGIS and georeference it by associating a geographic reference to the image so that it's clear where the area is physically located in space.

A secondary objective is to overlay other shapefiles on top of the aerial image. In this exercise, you will import an aerial image of the area around the Alamo, located in downtown San Antonio, Texas, and overlay a street network. **1**

Exercise file locations

Chapter directions: Follow the exercise as it appears in this book

This exercise uses the following:

- A shapefile of Bexar County streets. This file was also used in chapter 8, "Address mapping." If you did not do the exercise in chapter 8, but would still like the file, go to that chapter and follow the instructions on how to download a street network from the census. The chapter will walk through how to download the 2009 Bexar County street network (where San Antonio is located). If you prefer, you can instead use the file provided on this book's accompanying CD.

- An address locator for Bexar County streets. In this exercise, we'll use the address locator created in chapter 8, steps 3 and 4 called "**This is my Address Locator**." If you did not create this file, you should create it before beginning this exercise (or use the one provided on the CD accompanying this book).

- An aerial image of the area around the Alamo. This image is provided on the CD accompanying this book and is courtesy of DigitalGlobe, one of the largest aerial photo providers.

CD: Use the CD included with this book

All files needed for this exercise are included on the book's CD. Files are organized by chapter.

Personal files: Use files you've gathered from other sources

To complete this exercise, you will need the following:

- A shapefile of the area where your image is located
- An aerial photograph

1 Add a shapefile

1. Open ArcGIS.
2. Click the Add Data icon ⊕ ▾ .
3. Either open Folder Connections or select the Connect to Folder icon and navigate to **tl_2009_48029_edges.shp**, which was downloaded in chapter 8. Always add the shapefile before adding the JPEG map.
4. Click the Add button.

2 Add image into ArcGIS

1. Click the Add Data icon again and navigate to the aerial photograph. Either use the drop-down menu to navigate to the files (under Folder Connections) or select the Connect to Folder icon and navigate to the file.
2. When you get a dialog box that warns you of an Unknown Spatial Reference, click OK.

---

1 | **Georeferenced Aerial Photo and Street Shapefile The Alamo, San Antonio, Texas, 2009**

0   0.0125   0.025   0.05 Miles

Map Source: New Urban Research, Inc. 2009.
Image Source: DigitalGlobe, 2009.
Street Shapefile: U.S. Census TigerLines, 2009.

3. The image should be added to the table of contents, but will not be visible in Data View yet.

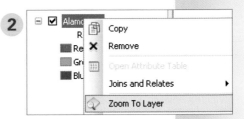

3 Move between the two layers

1. Even though you've added the image, it will not be immediately apparent. To view it, in the table of contents, right-click the aerial photograph's layer name and select Zoom to Layer. **2**

2. To view streets, do the same thing but with the streets layer.

    The idea is that we want both files to be displayed on top of one another. Right now the image file has no georeferencing, so the program has no idea where to put it in relation to the streets file. We'll fix this in the next few steps.

4 Visually identify four intersections in the aerial photograph

You must identify at least three points (the more the better, though) on the image that will serve like anchor points for georeferences. To make your task easier, four intersections have been identified and labeled in the illustration to the right. **3** ?

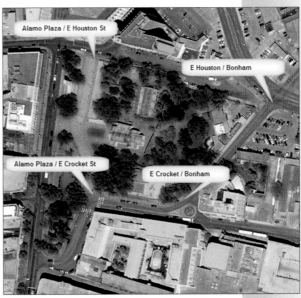

The four intersections identified here are

1. Alamo Plaza & E Houston Street (this intersection is in the northwest corner of the map)

2. Alamo Plaza & E Crockett Street (this intersection is in the southwest corner of the map)

3. E Crockett & Bonham (this intersection is in the southeast corner of the map)

4. E Houston & Bonham (this intersection is in the northeast corner of the map)

NOTE: This exercise uses an address locator for Bexar County streets. In this exercise, we'll use the address locator created in chapter 8, steps 3 and 4 called **This is my Address Locator**. If you did not create this file, you should create it before beginning the exercise (or use the one on the provided CD).

5 Identify the first intersection on the street network in ArcGIS

1. In ArcGIS, open the Find tool, which looks like a pair of binoculars, next to the Identify tool. 🔍 **4**

2. Select the Locations tab.

3. Click the icon next to Choose a locator: and navigate to **This is my Address Locator** created in chapter 8 (and also provided on the CD). **4**

4. Click Add.

5. In the Street or Intersection field, type **Alamo Plaza & E Houston St**.

   Here you must connect the intersections with a symbol that the address locator is expecting. There are three default intersection connectors built into the address locator: & | @.

6. Click Find.

7. Click the first intersection (which is usually the correct one) in the Address Description section. Notice that the spot on the map flashes, but the view is still too far out to see much. You may need to move the box over slightly so you can see the map and flashing point. Usually the first address given is the best match. This one had a score of 88 out of 100. Use the scroll bar to review the intersectional address. **5**

8. Right-click the first intersection and select Add Point. This will place a point at that intersection, so it's very easy to see.

CREATED ADDRESS LOCATOR VERSUS NORTH AMERICAN GEOCODE SERVICE (ARCGIS ONLINE)

For future reference, if you don't have an address locator, you are able use the option North American Geocode Service (ArcGIS Online). This is sort of an all purpose option. It works fairly well but you will get more accurate results using an address locator created specifically for this street network. Since we already have one created from chapter 8, we will use it in this exercise.

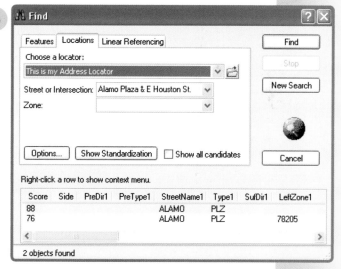

9. Use the Magnifying tool to zoom in to see the placement of the dot. 🔍 **6**

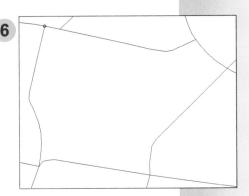

> **GOOD TO KNOW**
> You can also right-click the intersection and select Flash to flash the point on the screen so you can see where it is. This is helpful just to get a visual and confirm it's the right spot before you place the graphic.

6 Identify three other intersections on the street network in ArcGIS and place points

1. Do the same steps with the other three addresses. The remaining intersections are

- Alamo Plaza & E Crocket Street (For this intersection, the second option is the correct one. The only way to know that is to zoom in, use the Identify tool and do some guesswork, which has already been done for you.)

- E Crockett & Bonham

- E Houston & Bonham

2. After you find each intersection, in the Address Description section of the dialog box right-click the first intersection and select Add Point. This will place a point at that intersection. **7**

3. Select the Identify tool then click the streets where points have been positioned, to confirm that you are in the right area. ℹ️

4. Close the dialog box.

7 Copy and paste the longitude (X) and latitude (Y) and values of the four intersections

1. Click the default pointer to activate it. ⬉

2. Right-click on the Alamo Plaza & E Houston Street point (the northwestern most point) and select Properties. **8**

3. Select the Size and Position tab. The x,y values will be displayed. **9**

4. You need to copy and paste these x,y values. Open Microsoft Word, Excel, or Notepad to store the values.

5. Type the four intersections similar to the table below.

9 **Alamo Plaza & E Houston Street**

| Symbol | Location | Size and Position |

Position

X: |-98.486448 dd

Y: 29.426493 dd

Intersection	X values (longitude)	Y values (latitude)
Alamo Plaza and E Houston Street (this intersection is in the northwest corner of the map)	-98.486448 dd	29.426493 dd
Alamo Plaza and E Crocket Street (this intersection is in the southwest corner of the map)	-98.486516 dd	29.425064 dd
E Crocket and Bonham (this intersection is in the southeast corner of the map)	-98.485359 dd	29.424969 dd
E Houston and Bonham (this intersection is in the northeast corner of the map)	-98.484538 dd	29.426052 dd

6. Highlight the values in the x field, right-click, select Copy, switch to the software where you will store the x,y values, right-click, and select Paste.

   NOTE: Your numbers may be slightly different.

8 Turn on the Georeferencing toolbar and create control points on the image

Here's where things get a bit tricky. The idea is to identify the same four points on each layer (the JPEG and the shapefile). To do this, use the following steps:

1. On the Customize menu, select Toolbars and Georeferencing. Dock the toolbar by dragging it to the top of the window.

2. In the table of contents, right-click the aerial photo's name (**Alamo.jpg**) and select Zoom to Layer. Your aerial should appear.

3. Click the View Link Table tool on the Georeferencing toolbar. ⊞

4. Click the Add Control Points tool on the Georeferencing toolbar. ⤢

CONTROL POINTS

The x and y values you are collecting here will ultimately be used to align the JPEG image to the shapefile. The more control points you have, the better fit you will get. The minimum is three.

5. On the JPEG image, at the first intersection (Alamo Plaza and E Houston Street), click once in exactly the same spot on the aerial photo, then click again (a slow double-click) where the centerlines of the two streets meet. A plus-sign symbol at the control point turns from green to red when it's placed, and numbers are filled in the link table. **10**

6. The link table should automatically fill in the appropriate values for X Source, Y Source, X map, Y map, and residual values. The *X source* and *Y source* are the coordinates on the *image*. The *X map* and *Y map* values will be the corresponding coordinates on the street network *shapefile* (which we'll fix in the next step).

7. Use the same procedure for the other three intersections, noting the order in which you select them as you'll need to update the X map and Y map values in the same order in the next step.

CAN'T SEE WHERE THE INTERSECTIONS ARE LOCATED?

If you're having a difficult time figuring out where the intersection is while looking at the JPEG map in ArcGIS, you can use all the standard tools to orient yourself, including the Information tool. It won't mess anything up.

9 Input corresponding values for the shapefile

1. Remember that the *X source* and *Y source* are the coordinates on the *image*. The *X map* and *Y map* values will correspond to the coordinates on the street network *shapefile*.

2. Click the first record in the Link Table and enter the corresponding X map and Y map values that you copied and pasted earlier. Copy and paste over the values that are currently there. Be sure to do this in the correct order so you are matching the correct intersection with the corresponding correct intersection.

3. After doing one or two points, things may look a little crazy (you may also get some error messages; ignore them) and you may not be able to see either layer. In the table of contents, right-click the image layer name and select zoom to layer. You should start to see things shaping up.

4. Enter the X map and Y map values for the other three records. If you make a mistake it's not a big deal. Simply highlight that record in the List Values table and Click the icon to delete that entry and just do it again. You should start to see a gradual alignment of the JPEG image and the shapefile.

5. Once you're satisfied with the results, click OK.

10 Check accuracy

You can evaluate the overall quality of the work using the measure Total RMS Error.

1. Open the Link Table by clicking the View Link Table icon.

2. In the lower right corner, notice the Total RMS Errors box. If your total RMS error is bigger than 5 meters, delete the link with the highest residual in the Link Table. Go back to the image and reenter the control point. After you have gotten an acceptable total RMS error, click OK.

11 Update georeferencing

1. Once you've finished adding points and working with the data, select Georeferencing on the Georeferencing toolbar.

2. Click Update Georeferencing. Now the image can be used in conjunction with other shapefiles for this area. **11**

TRANSFORM AND ADJUST

On the Georeferencing toolbar you can select Transformation and Adjust. In most instances, this will better align the control points on the image and the shapefile. Try it now and see if it helps. If not, you can always undo the adjustment.

ACCURACY

After you get the two images overlaid, notice there are still some errors between the aerial photograph and the street layer. These are to be expected. It's not a perfect science. The goal is to get two images to match as best they can, given your time and patience for doing this type of work. Adding more control points would help.

If your project calls for extreme accuracy, you would need to repeat this process on very small sections, one at a time, then merge all the pieces together.

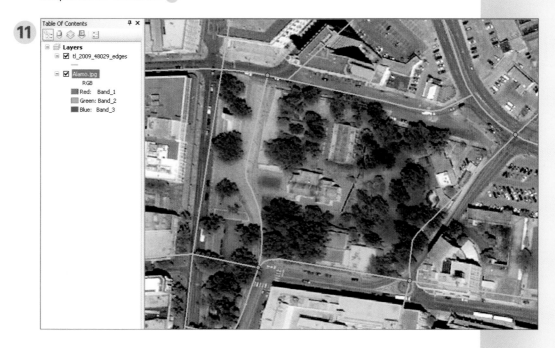

CHAPTER 14

Digitizing paper maps

Many organizations still work with paper maps but would probably prefer working with digital versions of these maps since they are easier to update. It isn't hard to create a digital version of paper maps. It's also very helpful to be able to overlay other shapefile boundaries on top.

First you need to scan a map using a scanner and save it as a JP EG file. PDFs will not work. If the map is in PDF format, you need to convert it to a JPEG. If your map is too large to scan, you need a digitizing tablet or pen specially created for this purpose.

This exercise is similar to the chapter 13 exercise. However, instead of bringing in aerial photography, we'll import a scanned JPEG image of the Providence Community Housing target area in the Treme neighborhood of New Orleans. Then we'll draw a polygon boundary and create a new shapefile. We'll use the target area boundary as a visual guide to draw our new boundary.

Exercise goal

Import and georeference a scanned map into ArcGIS, then create an empty shapefile, which represents a target area boundary. **(1)**

Exercise file locations

Chapter directions: Follow the exercise as it appears in this book

This exercise uses the following:

- A shapefile of Orleans Parish streets (where New Orleans is located). This file is *not* downloaded during this exercise. Steps on how to download shapefiles from the U.S. Census are included in chapter 8. Follow the instructions in steps 1 and 2 to download a 2009 street network, except select Orleans Parish, Louisiana, instead of Bexar County, Texas. If you prefer, you can instead use the file provided on this book's accompanying CD.

- A scanned JPEG image of the Providence Community Housing target area, located in the Treme neighborhood, Ninth Ward, New Orleans, Louisiana. This file is provided on the book's CD, courtesy of Providence Community Housing.

CD: Use the CD included with this book

All files needed for this exercise are included on the book's CD. Files are organized by chapter.

Personal files: Use files you've gathered from other sources

- To complete this exercise, you will need a street network of where the area is located.

- You will also need a JPEG of the area you want to work with in ArcGIS.

1 Add shapefile

1. Open ArcGIS.

2. Click the Add Data icon ⊕ ▾ .

3. Either open Folder Connections or select the Connect to Folder icon and navigate to **tl_2009_22071_edges.shp**, the street network for New Orleans. The streets may look distorted, but that's okay. Always add the shapefile prior to adding the JPEG map.

4. Click Add.

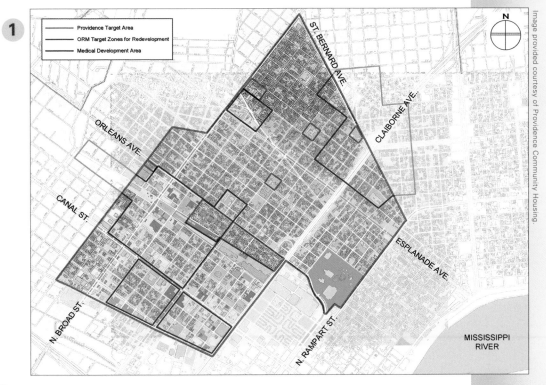

Image provided courtesy of Providence Community Housing.

2 Add the scanned JPEG image and build pyramids

1. Click the Add Data icon again.

2. Navigate to **TargetArea .jpg**, which represents the Providence target area in New Orleans (outlined in red).

3. Select the file and click Add.

4. After you add the JPEG image, you may be asked if you would like to create pyramids. Building pyramids allows your JPEG image to display more quickly. You may not always get this box depending on what your JPEG looks like. Click Yes to create pyramids. You won't see both layers together yet. **2**

5. When you are warned about an unknown spatial reference, click OK. **3**

6. The image will be added to the table of contents.

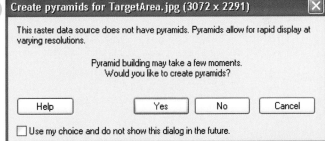

Create pyramids for TargetArea.jpg (3072 x 2291)

This raster data source does not have pyramids. Pyramids allow for rapid display at varying resolutions.

Pyramid building may take a few moments.
Would you like to create pyramids?

Help   Yes   No   Cancel

☐ Use my choice and do not show this dialog in the future.

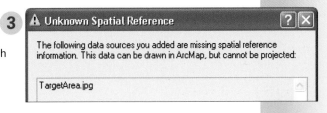

⚠ Unknown Spatial Reference

The following data sources you added are missing spatial reference information. This data can be drawn in ArcMap, but cannot be projected:

TargetArea.jpg

3 Open the Georeferencing toolbar and fit to display

1. On the Customize menu, select Toolbars and Georeferencing. Dock the toolbar by dragging it to the top of the window.

2. On the Georeferencing toolbar, elect Georeferencing then select Fit to Display, to display the raster image in the same general area as the shapefile. At this point, your map will look very strange. 4

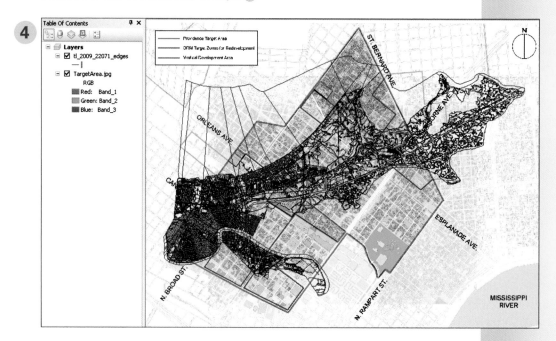

Turn each of the layers off and on and familiarize yourself with both files. The streets are on top of the image, but clearly it's not a good fit yet.

3. Move the streets layer to the top of the table of contents if it isn't already. Change the color of streets to black so you can see them clearly. Zoom out to get a broader perspective of both layers together.

4. Notice where N Broad and Canal streets intersect.

4 Find intersection 1 on the street map and place a point

Next we will add some points to help align the two images. To start, we have to find the first point to align.

1. In ArcGIS open the Find tool, which looks like a pair of binoculars, next to the Identify tool. 🔍

2. Select the Locations tab.

3. Since we haven't worked with this street network file previously, you would have no reason to have created an address locator. So for this exercise we'll use the North America Geocode Service (ArcGIS Online) option (unlike in the

previous exercise where we used a precreated address locator). **5**

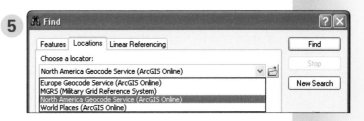

4. Type **N Broad St & Canal St** as the address, the City as **New Orleans** and **Louisiana** as the State/Province. **6**

5. Click Find.

6. Select the first intersection (which is usually the correct one) in the Address Description section. Notice the spot on the map flashes, but the view is still too far out to see much. You may need to move the box over slightly so you can see the map and flashing point. Usually the first address given is the best match.

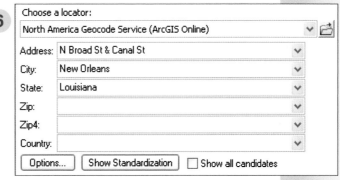

7. In the Address Description section of the dialog box, right-click the first intersection and select Add Point. This will place a point at that intersection, so it's very easy to see. **7**

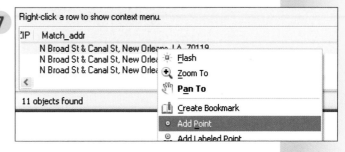

8. Close the Find dialog box.

9. Use the Magnifying tool to zoom in to see the placement of the dot.

ALTERNATIVE METHOD

You can also use the link table method outlined in chapter 13, "Aerial photography."

GOOD TO KNOW

You can also right-click the intersection and select Flash to flash the point on the screen so you can see where it is. This helps us see the point and confirm it's in the right spot before placing the graphic.

5 Add control point 1: N Broad Street and Canal Street

Adding control points allows you to align the JPEG with the shapefile layer.

1. Print the scanned map image now. It will make it much easier to see what you are supposed to do.

2. Look at the scanned map and find the intersection of N Broad Street and Canal Street (lower left side of map).

3. Click the Add Control Points tool on the Georeferencing toolbar. ✈ 　**8**

4. In the table of contents, turn off the street network to more clearly see the JPEG. You should still be able to see the dot (where the intersection occurs in the shapefile).

5. With the Add Control Points tool activated, click the place where the intersection is *on the JPEG first.* A line will appear. Drag that line to the dot and click the dot. *It must be done in this order. Click the JPEG first, then the shapefile.* The map should realign so the dot is in fact at the right place on the JPEG. This is control point 1. **8**

> **CONTROL POINTS**
>
> The more control points you have, the better fit you will get. The minimum is three, using four or five will usually do.

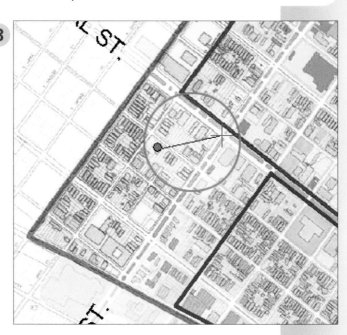

6 Find intersection 2 on the street map and place a graphic

1. Now do the same thing with St. Bernard Street and Claiborne Avenue on the right side of the map to create control point 2. If it's not already open, select the Find tool and type **Saint Bernard Ave & Claiborne Ave** in the street name section (the city and state should still be filled in).

2. Click Find.

3. Select the first intersection (which is usually the correct one) in the Address Description section. Notice the spot on the map flashes but the view is still too far out to see much. You may need to move the box over slightly so you can see the map and flashing point. Usually the first address given is the best match.

4. Right-click the first intersection in the Address Description section of the dialog box, and select Add Graphic. This will place a point at that intersection, so it's very easy to see.

5. Close the Find dialog box.

6. If necessary, use the Magnifying tool to zoom in to see the placement of the dot.

7 Add control point 2: St. Bernard Street and Claiborne Avenue

1. Look at the scanned map and find the intersection of St. Bernard Street and Claiborne Avenue on the right side of map.

2. Click the Add Control Points tool on the Georeferencing toolbar.

3. In the table of contents, turn off the street network so you can more clearly see the JPEG. You should still be able to see the dot (where the intersection occurs in the shapefile).

4. With the Add Control Points tool activated, click the place where the intersection is *on the JPEG first*. A line will appear. Drag that line to the dot and click the dot. *Again, it must be done in this order. Click the JPEG first, then the shapefile.* The map should realign so the dot is in fact at the right place on the JPEG. This is control point 2.

5. The entire image will get very small. Use the magnifying tool to zoom in. Turn the street layer back on and notice the street file and the image are beginning to line up, although it's not a perfect fit yet.

IF YOU MAKE A MISTAKE

Deleting a control point: If you accidentally add a point that you didn't mean to or that doesn't work well, simply delete it by using the View Link Table icon located next to the Add Control Points button.

8 Find intersection 3 on the street map and place a graphic

1. Now do the same thing with Esplanade Ave and N Rampart Street to create control point 3. If it's not already open, click the Find tool and type **Esplanade Ave and N Rampart Street** in the street name section (the city and state should still be filled in).

2. Click Find.

3. Click the first intersection (which is usually the correct one) in the Address Description section. Notice the spot on the map flashes, but the view is still too far out to see much. You may need to move the box over slightly so you can see the map and flashing point. Usually the first address given is the best match.

4. In the Address Description section of the dialog box right-click the first intersection, and select Add Graphic. This will place a point at that intersection, so it's very easy to see.

5. Close the Find dialog box.

6. Turn off the street network layer in the table of contents (if it's turned on).

7. If needed, use the Magnifying tool to zoom in to see the placement of the dot.

9 Add control point 3: Esplanade Ave and N Rampart Street

1. Look at the scanned map and find the intersection of Esplanade Ave and N Rampart Street (lower right side of map).

2. On the Georeferencing toolbar, click the Add Control Points tool.

3. In the table of contents, turn off the street network so you can more clearly see the JPEG. You should still be able to see the dot (where the intersection occurs in the shapefile).

4. With the Add Control Points tool activated, click the place where the intersection is *on the JPEG first*. A line will appear. Drag that line to the dot and click the dot. The map should realign so the dot is in fact at the right place on the JPEG. This is control point 3.

5. Turn the street layer back on and notice the street file and the image are beginning to line up, although it's not a perfect fit yet.

10 Find intersection 4 on the street map and place a graphic

1. Now do the same thing with Orleans Ave and Claiborne Ave to create control point 4. If it's not already open, click the Find tool and type **Orleans Ave and Claiborne Ave** in the street name section (the city and state should still be filled in).

2. Click Find.

3. Click the first intersection (which is usually the correct one). Notice the spot on the map flashes, but the view is still too far out to see much. You may need to move the box over slightly so you can see the map and flashing point. Usually the first address given is the best match.

4. Right-click the first intersection and select Add Graphic. This will place a point at that intersection, so it's very easy to see.

5. Close the Find dialog box.

6. Turn the street network layer off in the table of contents (if it's turned on).

7. If needed, use the Magnifying tool to zoom in to see the placement of the dot.

11 Add control point 4: Orleans Ave and Claiborne Ave

1. Look at the scanned map and find the intersection of Orleans Ave and Claiborne Ave (bottom of the map).

2. Click the Add Control Points tool on the Georeferencing toolbar.

3. In the table of contents, turn off the street network so you can more clearly see the JPEG. You should still be able to see the dot (where the intersection occurs in the shapefile).

4. For this last control point, you may want to zoom in super close so you can get a really good fit.

5. With the Add Control Points tool activated, click the place where the intersection is *on the JPEG first*. A line will appear. Drag that line to the dot and click the dot. The map should realign so the dot is at the right place on the JPEG. This is control point 4.

6. Turn the street layer back on and notice the street file and the image are fairly closely aligned. If you wanted an even better fit, you would add more control points. **9**

   Use the Identify tool and click the streets shapefile to verify the street names. ℹ️ **9**

12 Check accuracy

You can evaluate the overall quality of the work with the measure Total RMS Error.

1. On the Georeferencing toolbar, select the View Link Table tool.

2. In the lower right corner, notice the Total RMS Errors box. If the total RMS error is greater than 5, delete the link with the highest residual in the link table. Go back to the image and reenter the control point. After you have gotten an acceptable total RMS error, click OK in the Link Table dialog box.

13 Update georeferencing

1. Once you're finished adding points and working with the data, click Georeferencing on the Georeferencing toolbar.

2. Select Update Georeferencing.

Drawing boundaries by hand using a scanned map as a guide

In the scanned image, the red outline represents the Providence Community Housing target area. You may want a shapefile of the red boundary if, for example, it changes from year to year. You can use the JPEG map to draw a boundary by hand, as we'll do in the next section.

The next few steps assume that you have just georeferenced the scanned map image and have that image open. You can remove the street network file by right-clicking the street layer in the table of contents, and selecting Remove. You will not use this layer again, so it's good to get it out of the way.

14 Create a polygon shapefile

1. Click the ArcToolbox icon 📦 to open ArcToolbox.

2. Expand the Data Management Tool menu, then expand the Feature Class menu.

3. Double-click Create Feature Class.

4. In the first field Feature Class Location, click the browse folder icon and navigate to the folder where you will save what will be the newly created shapefile. Click the folder, then click Add.

5. In the second field Feature Class Name, type **NewTargetArea** (avoid spaces in file names).

6. Next to the Coordinate Systems field, click the Coordinate System button. 📄

7. Click the Select button.

8. Click the Geographic Coordinate Systems folder.

9. Select North America.

10. Double-click NAD 1983.prj.

11. Click OK twice.

12. **NewTargetArea.shp** should be added to the table of contents.

15 Make the layer editable

1. Click the Editor Toolbar button. In ArcGIS 10 this button should already be visible as it's a part of the default ArcGIS interface. In older versions of the software, you must go to View, select Toolbars and Editor. The icon will look the same. 🖼

2. Once you click the Editing icon, a new Editor toolbar should be visible. Dock the toolbar if it is not already docked.

3. Select Editor then Start Editing.

4. Notice a new Create Features box is added to the right.

16 Sketch a boundary

1. The concept here is that we are going to sketch out the boundary by placing dots (these are called vertices for the plural and vertex for the singular). Think of these like little dots that, once connected, create an outline.

2. At the bottom of the Create Features box, you'll notice a Construction Tools section. Select Polygon. This activates the Polygon tool, which is similar to the Sketch tool in earlier versions of ArcGIS.

3. To place the vertices, simply click along the red boundary line with the polygon tool (which will look like a plus sign). Notice that it places little dots wherever you click. Where there is a lot of detail in the map, you will need to place several vertices to outline it accurately. 10

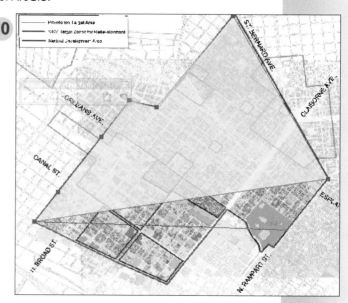

4. When you place the last one, completing the polygon, double-click and the boundary should show up with a bright blue outline color.

5. Click the Editor button again and select Save Edits, then Stop Editing.

17 Add an attributes column

1. In the table of contents, right-click **NewTargetArea**.

2. Open the attributes table.

3. Click the drop-down arrow for the Table Options tool on the Table toolbar (the first icon) and select Add Field. 🖼 ▾

4. Type in the name of the new column. Let's call it Area. For the Type, select Text.

5. Click OK.

6. Before you can type the Target Area name (**Providence Community Housing target area**) you must make the table editable. Click the Editor Toolbar button. (In ArcGIS 10 this button should already be visible as it's a part of the default ArcGIS interface. In older versions of the software, you must go to View, select Toolbars and Editor. The icon will look the same.)

7. Once you click the Editing icon, a new Editor toolbar should be visible. Dock the toolbar if it is not already docked.

8. Click Editor and select Start Editing.

9. In the new attributes field, type the target area name.

10. Click Editor again, select Save Edits, and then Stop Editing.

CHAPTER 15

Attribute queries

The ability to query data in an attributes table separates GIS software from graphic design software. Having the ability to query subselections of data allows users to analyze communities and problems in more sophisticated ways, to use the software to ask questions, and to perform intelligent analysis based on the answers.

The two most frequently used queries are attribute and location queries (graphics queries also exist and are used least often). An attribute query is simply a data query. Location queries are covered in the next chapter.

Exercise goal

Create an attribute query that identifies counties that have greater than 15 percent senior citizen population (derived in chapter 5). **1**

Exercise file locations

Chapter directions: Follow the exercise as it appears in this book
AgeJoined.shp, created in chapter 5. Complete the chapter 5 exercise if you did not already.

CD: Use the CD included with this book
All files needed for this exercise are included on the book's CD. Files are organized by chapter.

Personal files: Use files you've gathered from other sources
To complete this exercise, you will need a shapefile that has some data in it that can be queried using greater than or less than operators. Demographic information is a good choice. For example, you could query things like counties that have a population over 20,000 but less than 50,000 people.

1 Add a shapefile

1. Open ArcGIS.
2. Click the Add Data icon ✛ ▾ .
3. Either open Folder Connections or select the Connect to Folder icon and navigate to **AgeJoined.shp** that you created in chapter 5.

2 Write a query

1. From the Selection menu, choose Select by Attributes.

1 **Alabama Counties with 15% or Greater Senior Population, 2000** *Based on an attribute query*

2. Type the query as you see it in the graphic. For the query to work for counties that are greater than 15 percent, you must type in **.15**, not 15. ②

3. The counties that meet the condition are highlighted with a bright blue line.

4. Open the attributes table and notice that the corresponding rows are also highlighted.

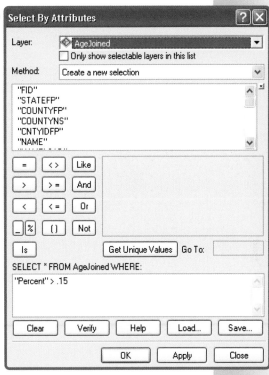

3 Create a new shapefile of selected areas

Notice that only counties meeting our requirement are selected. Let's make our query into a new shapefile that contains only the counties that answer the query.

1. In the table of contents, right-click **AgeJoined** and select Data, then Export Data.

2. Leave the default Export Selected Features and navigate to where you would like to save the file. Name the new file **Seniors**, and click Save and OK.

3. When prompted whether you would like to add this file to your current view, click Yes. Notice how the shapefile is added to the table of contents.

4 Export records (optional)

You can export the selected records from the attributes table.

1. In the table of contents, right-click **AgeJoined**.

2. Open the attributes table.

3. Click the Table Options tool arrow on the Table toolbar (the first icon) and select Export.

4. Navigate to where you would like to save the file. Give it a new name if you like. Notice that the file extension is .dbf. This is a generic database file type that can be read in Excel and Access.

5. To erase the query, on the Selection menu, select Clear Selected Features.

CHAPTER 16

Location queries

Location queries differ from attribute queries in that they don't involve selecting data. A location query is a geography query, not a data query. A location query involves selecting geographies within other geographies. It works equally well with points, lines, or polygons.

Exercise goals

Create a location query that selects cities within Tallapoosa County, Alabama, and then create a new shapefile. **1**

Exercise file locations

Chapter directions: Follow the exercise as it appears in this book

Files for this exercise were used in chapter 1:

- Alabama County and Equivalent (Current) shapefile
- Alabama Places (Current) shapefile

How to download these files is outlined in chapter 1, steps 1-3.

CD: Use the CD included with this book

All files needed for this exercise are included on the book's CD. Files are organized by chapter.

Personal files: Use files you've gathered from other sources

To complete this exercise, you will need at least two shapefiles of the same area but different geographic types. This might include cities and counties, counties and states, or streets and cities.

1 Add a shapefile

1. Open ArcGIS.
2. Click the Add Data icon ✛ ▾.
3. Either open Folder Connections or select the Connect to Folder icon and navigate to the Alabama counties shapefile used in chapter 1 (**tl_2009_01_county**).

2 Create a new shapefile for Tallapoosa County

The goal is to select those places that are in Tallapoosa County. First, we need a shapefile that contains Tallapoosa County only.

1. In the table of contents, right-click the county shapefile layer and open the attributes table and find Tallapoosa County in the list.
2. Click the gray cell at the beginning of the row. The row should become bright blue and also be highlighted on the map.
3. Close the attributes table.
4. In the table of contents, right-click the layer name, select Data, then Export Data. **2**

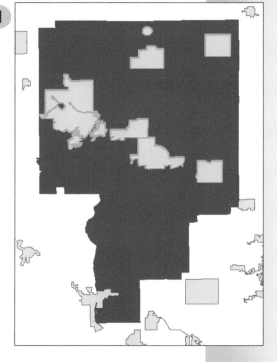

5. For the Export field, choose Selected Features (meaning just the highlighted Tallapoosa County).

6. Navigate to a where you would like to save the new shapefile on your C drive.

7. Name it **Tallapoosa**.

8. If it is not already selected, choose Shapefile as the Save As Type. Save as a Shapefile and click OK.

9. When asked if you would like to add to the map as a layer, click Yes.

10. Remove the shapefile with all the counties by right-clicking and selecting remove. Keep only the Tallapoosa County shapefile.

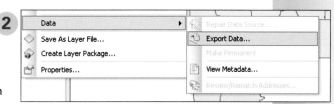

3 Create a location query

1 Use the Add Data icon again to add the Alabama places shapefile (**tl_2009_01_place**) to the map. This is the same file used in chapter 1.

2. On the Selection menu, choose Select by Location.

3. For the section underneath where it says Target Layer(s), check the check box next to **tl_2009_01_place**.

4. For the source layer, select the Tallapoosa shapefile.

5. For the "Spatial Selection Method" field, select Target Layer(s) features have their centroid in the source layer feature.

6. Click OK. 3

7. Notice that only those places within Tallapoosa county are selected. ⑦

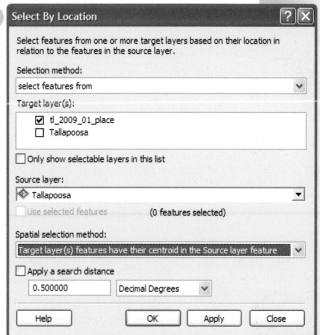

WHAT DOES "HAVE THEIR CENTROID IN" MEAN?

A centroid is the physical center point of geometry (in this case a polygon). By using the Have their centroid in method, the physical center of the place must fall within the county boundary, not merely touch it or intersect it. This method is more conservative. You will never get polygons that are only slightly within the boundary because the center point of the polygon must fall within the target boundary.

Another popular tool is Intersect, which would include any place that intersects the county boundary.

CHAPTER 17

Creating reports

If you use GIS to provide information to other people, you will probably at some point create reports to accompany maps. Reports can give your reader much more information and provide credibility for the map's data. Prior to ArcGIS 10, most reports were created outside of ArcGIS, although Crystal Reports was occasionally used. In ArcGIS 10, Crystal Reports has been removed from ArcGIS, and in its place is a new and improved report builder. Report data can either be included as a part of your map's layout (if it's a small amount of data) or attached as a technical addendum.

Exercise goal

Create and export a basic data report in ArcGIS. You will not be able to do this exercise if you are using a software version older than ArcGIS 10.

Exercise file locations

Chapter directions: Follow the exercise as it appears in this book

You created the file for this exercise in chapter 5: **AgeJoined.shp**

If you need to know how to create this file, go to chapter 5.

CD: Use the CD included with this book

All files needed for this exercise are included on the book's CD. Files are organized by chapter.

Personal files: Use files you've gathered from other sources

To complete this exercise, you will need at least one shapefile for which you would like to create a report.

1 Add a shapefile

1. Open ArcGIS.
2. Click the Add Data icon ✛ ▾ .
3. Either open Folder Connections or select the Connect to Folder icon and navigate to **AgeJoined.shp**, which you created in chapter 5.

   You may want to open ArcCatalog, create a new folder for this exercise, and copy and paste the file to the new folder (see chapter 20, steps 1 and 2, for working with ArcCatalog).

2 Open the attributes table

1. To refresh your memory about the types of data we have in this shapefile, in the table of contents, right-click **AgeJoined** and select Open Attributes Table.
2. Scroll to the far right and notice that for each county, you can see columns for total population, total number of seniors, and the percentage of seniors in the population. These columns will be the basis for our report.

3 Create a report

1. Click the Table Options arrow on the Table toolbar (the first icon) and select Reports then Create Report. **1**
2. Under Available Fields scroll down and select County. Click the right arrow to deposit it into the Report Fields section. Do the same thing with the Percent column. The report will have two columns. **2**
3. Click Next twice.
4. When you get to the menu to sort fields, select the drop-down for Fields and select the Percent column. Click the button next to it to sort the field descending. This will put higher percentages at the top of the column.

5. On this same menu, click Summary Options .

6. Summary Options calculates the average, which often is very helpful. It also does a few other calculations. Check the check box for the Avg option and click OK, then Next.

7. The next menu relates to how to layout the report. Leave the defaults and click Next.

8. At the Style menu, review the styles. Good options here are New York or Simple. Select New York and click Next.

9. Type **Percentage of Seniors** as the title of the report and then click Finish. Your data should display in report format.

4 Export report

Because the report is long (two pages), it is not practical to insert it into a layout. For this type of report, you may want to export it and attach to your map, or export it to Excel for further analysis.

1. On the Report toolbar, Click the Export Report to File icon

2. Click the Export Format arrow and notice several options. Select Portable Document Format (PDF).

3. In the File Name field, click the ellipsis and navigate to where you would like to save the file. Name it **SeniorReport**. Click Save and OK. This has saved your report in PDF format.

4. Close the report to return to ArcGIS.

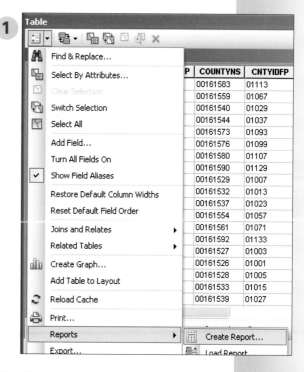

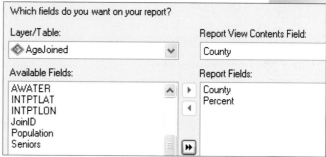

ATTRIBUTE QUERIES IN REPORTS

A useful report feature is the ability to write an attribute query and have the results of that query make up the report. When you begin to write a report, at the first page where you select the columns to include, if you click Dataset Options, you can select only certain records to include in the report. You have another option to select Definition Query as the option and a query wizard will open. Here you can write an attribute query that will select records based on your own criteria. See chapter 15 for more on writing attribute queries.

CHAPTER 18

Creating buffers

A buffer is a map item that represents a uniform distance around a feature (point, line, or polygon). Buffering is commonly used to pinpoint particular areas when doing spatial analysis. When creating a buffer, the user selects the feature to buffer around, as well as the distance of the buffer.

Exercise goal

Make 5-mile buffers around Alabama cities.

Exercise file locations

Chapter directions: Follow the exercise as it appears in this book

Place (Current) for Alabama state. You downloaded this file in chapter 1.

CD: Use the CD included with this book

All files needed for this exercise are included on the book's CD. Files are organized by chapter.

Personal files: Use files you've gathered from other sources.

To complete this exercise, you will need at least one shapefile for which you would like to make buffers. The file type can be point, line, or polygon.

1 Add a shapefile

1. Open ArcGIS.

2. Click the Add Data icon. ✛ ▾

3. Either open Folder Connections or select the Connect to Folder icon and navigate to **tl_2009_01_ place**, which was downloaded in chapter 1.

2 Create buffers

1. Click the ArcToolbox icon 🗔 to open ArcToolbox.

2. Expand the Analysis Tool menu, then expand the Proximity menu.

3. Double-click Buffer.

4. In the Input Features field, select the layer to buffer. In this case, choose the Alabama places shapefile (**tl_2009_01_ place**) and click Add.

5. In the Output Feature Class field, navigate to where you want to save what will be the newly created buffer file. Give it a new name such as **5milebuffer.shp** and click Save.

6. For the Linear Unit, type the buffer distance — in this case **5**. Select Miles.

7. Leave all other default options and click OK. A new buffer shapefile will be created. **1**

OVERLAPPING BUFFERS

To dissolve barriers between any overlapping buffers, click Dissolve Type and select LIST.

3 Rearrange shapefiles and review attributes table

1. To move the places shapefile to the first position in the table of contents, you must click the List by Drawing Order icon in the table of contents. **2**

2. In the table of contents, drag the places shapefile to first position.

3. In the table of contents, right-click **5MileBuffer** and select Open Attributes Table. After a buffer has been created, the buffer distance is added to the new layer's attribute table.

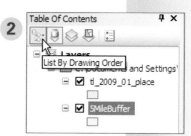

CHAPTER 19

Publishing maps

You have several options if you want to publish and share maps, including ArcGIS Explorer, a free GIS viewer that is available to download online. Using ArcGIS Explorer provides a good way to share maps and geopresentations with others. ArcGIS Explorer is similar to Google Earth—it's fairly intuitive and easy to use for even the most inexperienced map reviewer. All the information is contained in one presentation that you can e-mail or upload to a Web site. All a reader needs to view the presentation is ArcGIS Explorer.

With the release of ArcGIS 10 came the ability to create map books using data-driven pages. This new functionality is helpful if you need to create a book of maps where the header placement, legend, and other map elements are static, but the maps change. The final product is a series of geoenabled PDFs. If you use newer versions of the free Adobe Reader software to view your PDF maps, you are able to get correct latitude and longitude points as well as see a dialog box with attribute data. How cool is that?

Exercise goals

Create a geopresentation in ArcGIS Explorer, a singular geoenabled PDF, and a map book.

Exercise file locations

Chapter directions: Follow the exercise as it appears in this book

Files for this exercise were used in chapter 6. The files needed are the following:

- Seniors_Thematic.mxd
- ArcGIS Explorer
  - To download, go to **http://resources.esri.com/arcgisexplorer/1200/index.cfm?fa=home.**
  - Click the Download Now link in the upper left corner.
  - Select the language you wish to download in.
- Adobe Reader (free version).

If you do not have Adobe Reader 6.0 or higher, go to **http://get.adobe.com/reader/** and download it. Install Adobe Reader before beginning this exercise.

CD: Use the CD included with this book

All files needed for this exercise are included on the book's CD. Files are organized by chapter.

Personal files: Use files you've gathered from other sources

One or more shapefiles that contain some features in detail of a community. Ideally, a thematically mapped layer would be used.

- ArcGIS Explorer:
  - Go to **http://resources.esri.com/arcgisexplorer/1200/index.cfm?fa=home.** (**Note:** The English version of ArcGIS Explorer is included on the CD that came with this book.)
  - Click the Download Now link in the upper left corner.
  - Select the language you wish to download in.
- Adobe Reader (free version). If you do not have Adobe Reader 6.0 or higher, download it here **http://get.adobe.com/reader/**. Please install before beginning this exercise.

Publishing maps with ArcGIS Explorer

1 Open the project in ArcGIS

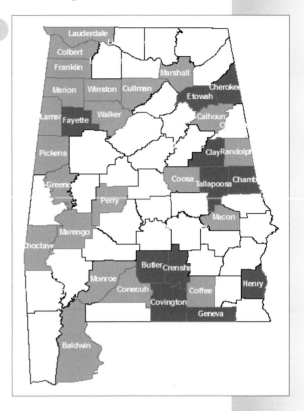

1. Open ArcGIS.

2. Select File, Open, and **Seniors_ Thematic.mxd**. (This project was created in chapter 6 and is also available on the book's CD).

3. Click Open. Your thematic map should be visible now. **1**

2 Create a layer package

1. In the table of contents, right-click the layer name and select Properties, and then select the HTML dialog box tab.

2. Select the first option to Show contents for this layer using the HTML Popup tool. Leave all other options as they are. This will allow all the attributes to be displayed in ArcGIS Explorer via a dialog box.

3. Select the General tab, and in the box beside Description write "Seniors Population in Alabama" and click OK. This step must be completed to validate the layer package.

4. In the table of contents, right-click the layer name and select Create Layer Package.

5. Click the button next to Save package to file, click the ellipsis and navigate to where you would like to save the layer package. You can leave the name AgeJoined, and the extension .lpk will be added. Click Save.

6. Click Validate and then Share.

7. Click OK.

8. Keep ArcGIS open because you'll need it later.

THE DIFFERENCE BETWEEN LAYERS AND LAYER PACKAGES

Layer packages were introduced with the 9.3.1 version of ArcGIS Desktop. A layer package is a single file that contains both a map layer and all the data the map uses. A layer only includes the map layer, but no associated data. Both maintain a map's color shading and labeling.

USING LAYER PACKAGES IN ARCGIS

In ArcGIS you can easily open a layer package in one of two ways. Either drag the layer package file and drop it into the ArcMap window, or right-click the layer package file and select Unpack.

3 Open ArcGIS Explorer and the layer package

NOTE: You must unzip and install ArcGIS Explorer before you can complete this step.

1. Open ArcGIS Explorer by double-clicking the ArcGIS Explorer icon on your desktop. If you do not see an icon, access the program from the Start Menu. Choose All Programs, ArcGIS Explorer, ArcGIS Explorer.
2. On the Home tab, click the Add Content icon, select ArcGIS Layers.
3. Navigate to, and then select, **AgeJoined.lpk**. Click Open.
4. To see the legend, click the Tools tab and select Show Legend (optional).

ArcGIS Explorer has many useful things to offer that are not covered in this chapter. Within the application, click the F1 key to access an extensive help menu.

4 Make the layer semitransparent (optional)

1. Switch to the Tools tab and click the Transparency button.
2. Slide the tool about half way up to make the layer semitransparent.

5 Add a note and Web site link to the map

1. Click the Home tab and Select Point.
2. On the map, click Birmingham.
3. Type **Largest City** as the note header. In the note section, type **http://en.wikipedia.org/wiki/Birmingham,_Alabama**. This is the link to Birmingham's Wikipedia page.
4. Click OK, and the page should display. Once you're finished looking at it, click the X in the upper right corner to close the page.

   NOTE: You must have an Internet connection.

5. Click the red note icon on the map to see how the Wikipedia page pops up.

   NOTE: You must have an Internet connection.

6. Close the note box. **2**

6 Link to Web sites and files

Another way to add Web site addresses to your map and presentation is to use the Add Link button. You can also link to any file to display to your audience. The ability to link to, and simply click, multiple Web sites and documents is very helpful during presentations.

1. Click the Home tab and the Link button.

2. Click the Browse arrow to navigate to Alamo.jpg used in chapter 13. For the purposes of this example, it doesn't really matter which file we link to, so let's use one that we've already worked with and have easy access to. It doesn't matter that it's a picture of the Alamo, it could be any document. This file can also be found on this book's accompanying CD. Click Open and then Create.

3. Change the file name to **Alamo Aerial** by clicking the file name once. The text will become editable so you can change the file name.

4. Double-click **Alamo Aerial** to display the file.

5. Now let's add a link to a Web site. Click the Link button again.

6. In the space provided type **http://en.wikipedia .org/wiki/Birmingham,_Alabama**. Click Create.

7. Change the file name to **Alabama Wiki**.

8. Double-click **Alabama Wiki** to display the Web site.

   NOTE: You must have an Internet connection.

7 Access attribute data for the layer

Since this is a layer package (versus a layer) all the data associated with the map is included. Click any county to see a dialog box with data related to that county. **3**

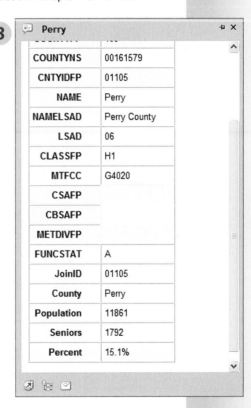

8 Prepare to create a presentation

1. Click the Display tab. In the Distance Units box, click the drop-down arrow and select Feet, Miles. The scale bar should now be displayed as miles.

2. Before we begin creating a presentation, you'll need to adjust the map to whatever size you deem necessary. In this example, zoom out to the point where all of Alabama is visible. Go to the Home tab, select the Zoom To button and select Country. Your map will begin to zoom out. When it looks similar to the image here, click the map to stop the zoom process. ④

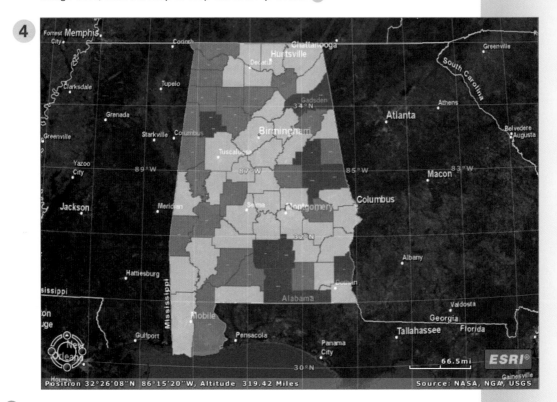

9 Create a presentation

1. To begin creating a presentation, click Edit Presentation on the Home tab. A new tab is opened called Presentation.

2. Click Edit Text to insert a title for the first slide. Type Distribution of Senior Population by Alabama County, 2000. Click OK. The font is likely too large, so click the Decrease Font Size button to make the title smaller. You want to see all text within the window.

3. You can also select the Quick Styles button to select a title template. Select the template that has a black background and yellow type.

4. Now we're going to capture our first slide. Click Capture New Slide. This button works like a camera taking a picture of whatever is on the screen.

5. Now zoom in to a point of interest. In this case zoom to one of the darkest shaded areas such as Fayette County on the northwestern side of the map. To zoom, you must switch back to the Home tab, select Zoom to the City. Once the map is suitably zoomed in, click it to stop the zoom process and drag the map so the area of interest is centered.

6. Click Edit Presentation again, then Capture New Slide. The second slide should be captured.

7. To capture a third slide, click Fayette County (or whichever county you selected) to generate the informational popup box. Scroll down slightly to reveal what the percentage of Seniors is for that county. Click Capture Slide button. Close the popup box.

8. Click the map and drag it to the left to reposition on Birmingham. Click Capture New Slide button.

9. Click the Largest City point created earlier. The Wikipedia page for Birmingham should open. Widen the box horizontally to take up much of the screen, while still giving you accessibility to the Capture New Slide button. Click Capture New Slide.

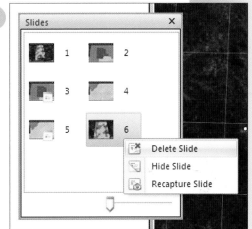

10. Close the Largest City notes box by clicking the X in the upper right corner.

11. On the Home tab, click the Zoom To button and zoom to the state level. Click Edit Presentation, and then Capture New Slide. You should now have a total of six slides. If you accidentally clicked a slide that you didn't mean to, you can right-click the slide and delete it. **5**

10 Set presentation options, play slide show, and save

1. On the Presentation tab, click Options.

2. In the Advanced Slides section, select the Automatically after option, and change from ten seconds to five seconds.

3. Click OK.

4. Click the Close Presentation Tab button.

DISPLAYING THE LEGEND

To display a legend, click the layer name in the Contents window, select the Tools tab, and click the Show Legend button. The legend will likely be fairly large. Adjust the size of the legend box by clicking the bottom of the legend and moving your mouse upward. This will shorten the legend. Move the legend to the lower left corner of the map.

5. Click the Start Presentation button to begin the presentation.

6. Once you're finished viewing the presentation, click the X button to return to the ArcGIS Explorer interface.

7. In the upper left corner, click the ArcGIS Explorer logo and select Save As and navigate to where you would like to save the project on your C drive. Give it the name **GeoPresentation**. Notice that it has the file extension .nmf, which indicates it is an ArcGIS Explorer file. Click Save.

8. Close ArcGIS Explorer.

> **SHARING PRESENTATIONS**
>
> To e-mail a presentation to colleagues, click the ArcGIS Explorer button and select Email, then Email Map. The ArcGIS Explorer file is sent in an e-mail with the following note:
>
> *The attachment to this e-mail can be viewed with ArcGIS Explorer. You can download and find information about ArcGIS Explorer at the ArcGIS Explorer Resource Center.*
>
> **http://resources.esri.com/arcgisexplorer**
>
> To begin the presentation, your reader clicks Start Presentation.

Publishing geoenabled PDF maps

Another good way to share your map is to export it as a geoenabled PDF. In Adobe Acrobat Reader 6.0 and higher, you can turn on and off individual map layers. With Adobe Reader 9.0 you'll be able to use many new tools that work with ArcGIS. The step below works with either version, but it's a good idea to upgrade if you haven't. It's free software.

11 Modify attributes table

Here we are going to select only a few columns to bring over with the PDF since we don't need all columns. Here's how you choose which ones to bring in:

1. In ArcGIS, open the Seniors_Thematic.mxd if it is not already open.

2. In the table of contents, right-click the layer name. Select Properties.

3. Select the Fields tab and notice that all columns have a check mark next to them. If you were to leave it like this, all these columns of data would be exported.

4. Select the Turn All Fields Off icon. ▯ **6**

5. Check the Name, Population, Seniors, and Percent check boxes. These are the only columns that will come with the PDF. Click OK.

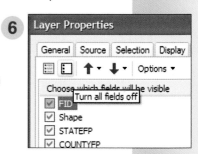

6. In the table of contents, right-click the layer and select Open Attributes Table. Notice that now you only see the selected columns.

7. Close the attributes table.

12 Export the map as a PDF

1. On the File menu, select Export Map.

2. Give the file a new name such as **AdvancedExport**.

3. For the Save As type, select PDF.

4. Select the Advanced tab.

5. For Layers and Attributes select Export PDF Layers and Feature Attributes. Place a check beside Export Map Georeference Information.

6. Click Save. **7**

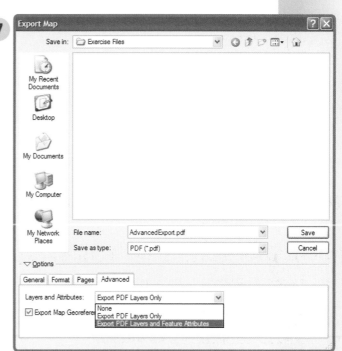

13 Look at layers in Adobe Reader

1. Double-click the file to open it in Adobe Reader.

2. In Adobe Reader, on the navigation menu in the left pane, select the Layers icon. It is the second option and looks like sheets of paper stacked on top of each other.

3. Click the plus sign in front of Layers to expand the menu. Click the second plus sign as well.

4. Although we only have one layer open here, click the little eye icons to turn off and on map elements. If you had multiple layers, you could turn off and on specific ones.

14 Use the Geospatial Location tool

1. A geospatial tool is included in the newest version of Adobe Reader. On the Tools menu, point to Analysis then Geospatial Location Tool. **8**

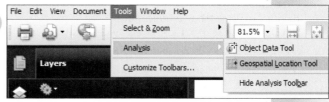

2. Move your cursor over the map and you'll see the latitude and longitude displayed.

15 Access attribute information from within the PDF

1. On the Tools menu, point to Analysis then the Object Data tool.

2. On the map double-click any county. It will become outlined in red. You may have to double-click twice.

3. Under the Model Tree, the attribute information for the selected county displays on the left at the bottom of the screen. **9**

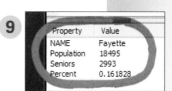

Creating map books

With ArcGIS 10 you can create maps using
functionality called data-driven pages. In the
next few steps we'll create a map book that
consists of one PDF page for each county. (?)

16 Create a map book of counties

1. When you create a map book, you cannot undo the book, so to be sure nothing
   gets messed up with our project, let's save a new project solely for the purposes
   of creating a map book. From the File menu, select Save As and navigate to
   where you would like to save your new map book project on your C drive.
   Name it **MapBook.mxd**. Click Save.

2. In the table of contents, right-click the layer name, and select Properties and
   then the Labels tab.

3. Change the color of the labels from white to black, and make the font 12-point.
   Click OK. Your map may look messy, but that's OK for now. It will soon be fixed.

4. From the View menu, select Layout View. It's best to create a map book from the
   Layout View.

5. On the Layout toolbar, click the Zoom to Whole Page icon on the Layout toolbar
   so you can see the full layout.

   NOTE: If your map isn't centered, feel free to reposition it using the Pan tool, but it's
   not necessary.

6. Using the default point, select the title and press the Delete key on your
   keyboard to delete the title. Also, select the source text and delete that as well.
   Only the map and legend should be included in the layout.

7. On the Customize menu, select Toolbars, then Data Driven Pages. A new toolbar
   will be added to the window.

8. Select the Page Text icon Page Text ▾ and then
   Data Driven Page Name.

9. A very small text box will appear in the
   center of the map. Drag the text box to the
   upper left corner of the map, right-click, and
   select Properties. This will allow you to have
   a title on each page with the county name.

10. In the attribute table there are two columns
    that would work well for page titles: NAME or
    NAMELSAD. NAME has just the county name
    without the word "County" behind it. The
    column NAMELSAD, however, contains both
    the county name and the word county behind it.
    Let's use the NAMELSAD column to create the
    titles. Right-click the title text box and replace
    Name with **NAMELSAD**. Also, click Change
    Symbol and make the font bold and size 22. **10**

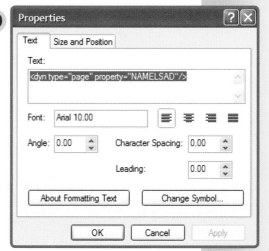

11. Click OK twice.

12. Click the Data Driven Page Setup icon. It might be difficult to see, but it's the first button on the new toolbar.

13. Select the Enable Data Driven Pages check box.

14. For the Layer Field, select AgeJoined.

15. For the Name Field, select Name. This is the column that determines what field will be used to create the various maps. We've selected Name, which is the name of each county, so each county gets a page. Leave all other default settings as they are. Click OK. Also change the Sort field to sort, so the pages will be alphabetically sorted by county name. Leave all other default settings as they are. Click OK.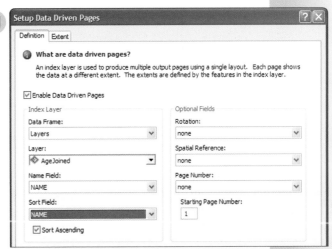

16. Review the results. Scroll through the pages of the new map book by clicking the blue arrows to scroll to the next page.

17 Create a PDF of the map book

1. On the File menu, select Export Map.

2. If the Save As type is not already .PDF, select .PDF from the menu.

3. Click the Pages tab.

4. This map book should have sixty-seven pages (one for each county). Let's only print pages one through five. Select the Page Range option and type **1-5**.

5. Click Save.

6. Open MapBook.pdf and review the results. These are geoenabled PDFs.

CREATING A MAP BOOK FROM A GRID

You may find it necessary to create a map book based on a grid, instead of by county like we used in this example. ArcGIS can create a grid with assigned numbers such as A1, A2 for each grid cell, and then create a book where each page corresponds to a grid cell. The tool to use is Grid Index Features to create a grid.

CORRECTING THE LEGEND

Occasionally, you may encounter an issue with the legend where you've already selected white as the legend background color, but you can still see through parts of the legend around the edge. The fix for this is to delete the legend and reinsert it, again selecting white as the background color.

KML AND KMZ FILES

KML stands for Keyhole Markup Language. KMZ stands for KML-Zipped. It is the default format for KML and it is a compressed version of the file. These file types are very useful for uploading maps to Google Earth and Google Maps. (Keyhole was the name of the application before Google bought it and added its own features and larger databases.) Many other Web applications also use KMZs.

It's easy to convert any layer to a KMZ. Search for the Layer to KML tool and export the layer as a KMZ.

Creating geodatabases

You may encounter many types of files while working in ArcGIS. We've been working with shapefiles, the most basic and frequently used file type in ArcGIS. Shapefiles are used extensively not only in ArcGIS but with many other GIS software programs. Another commonly used file type is the geodatabase.

A geodatabase is like a container where you can store all files related to your GIS project. You can store multiple shapefiles, aerial photographs, spreadsheets, and many other types of files. To group these items together in one place is very convenient since it provides one point of access for all needed files. It's also handy if you would like to share these files with others. Essentially, putting everything in a geodatabase makes GIS data easier to manage and access.

> **THREE TYPES OF GEODATABASES**
>
> ArcGIS supports three types of geodatabases: file (.gdb), personal (.mdb), and ArcSDE (.sde).
>
> For one person, or small workgroup projects, the file geodatabase is the best choice. It has no size limitations. The personal geodatabase is still used, but this is an older file type and has a 2 GB size limitation.
>
> The ArcSDE geodatabase (SDE stands for Spatial Database Engine) is for large workgroups. It is not accessible with a standard ArcView license. This type of geodatabase would be created and administered by a GIS manager. SDE has some big advantages. Files can be locked so they cannot be edited by different people, helping to maintain data integrity. Also, SDE can store and draw raster data, like aerial photographs, very efficiently. Another key advantage is that multiple people can work, and even edit data, simultaneously, unlike in file and personal geodatabases.

If you're only using a few shapefiles for your GIS project, there is no need to convert it to a geodatabase. However, if your project begins to grow and use many files, it is worthwhile to convert the project to a geodatabase. In any case, it's good to be familiar with geodatabases, as many organizations distribute files in this file format. Even though you may never find the need to create one, it is likely that at some point you may encounter one.

Exercise goal

Create and work with a file geodatabase using ArcCatalog.

Exercise file locations

You do not need any files to complete this exercise.

ArcCatalog is the library system of ArcGIS. You can browse, copy, delete, and organize files. ArcCatalog is also where you create file and personal geodatabases.

1 Open ArcGIS and ArcCatalog

1. Open ArcGIS.

2. On the Windows menu (at the top) select Catalog. (You can also click the Catalog icon).

   Notice that a new window has opened up on the right side of the screen. This is ArcCatalog. ArcCatalog looks slightly different in older versions of the software. **1**

2 Explore ArcCatalog (optional)

While you're here, you may as well learn a few things about ArcCatalog (if you're not already familiar with it).

1. Click the next Folder Connections link and notice that any saved connection to folders you may have previously set up are located in the box at the bottom.

2. Right-click Folder Connections and select Connect Folder. **2**

3. Double-click the C drive and navigate to a folder that you might use repeatedly (in this example, C:\Exercises), and click OK. In a few steps this is where you'll create a file geodatabase. You can also connect to a folder by clicking the Connect to Folder icon on the ArcCatalog toolbar.

4. Notice that now when you click that new link under Folder Connections, all the files in the folder are displayed and accessible in the box at the bottom.

5. Assuming you have a file available, right-click the file name and notice the menu of options here. You can create new folders as well as copy, delete, rename, and export files. This is very handy for managing your GIS files. **3**

3 Create a file geodatabase

The basic database process is to first create the database, then export files to the geodatabase.

1. In ArcCatalog, double-click the Folder Connections link and right-click the folder where you would like to save your file geodatabase (in this example, C:\Exercises).

> ### GOOD TO KNOW
> You can also drag and drop files from ArcCatalog into the table of contents to easily open files.

2. Select New and then File Geodatabase.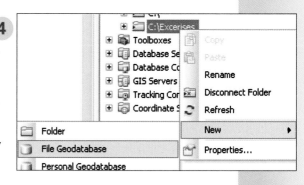

3. When prompted to give it a new name, call it **MyNewGDB.gdb**.

4 Add a shapefile to the geodatabase

1. In ArcCatalog, navigate to the street network (**tl_2009_48029_edges.shp**) used in chapter 13, "Aerial photographs." It's a street network for Bexar County, Texas. (Or you can access it via the CD provided with this book).

2. Right-click the file and select Export, and then select To Geodatabase (single).

3. The first field Input Features should already have the file you want to export to the geodatabase.

4. In the second field Output Location, use the browse button to navigate to the file **MyNewGDB.gdb**. You will likely have to use the drop-down menu to select Folder Connections and then navigate to the folder where you created the geodatabase. Select the geodatabase and click Add.

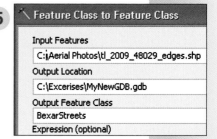

5. In the Output Feature Class field, type a new name for the file. Give it the name **BexarStreets**.

6. Click OK. It will take a minute to process.

7. Once it finishes, double-click **MyNewGDB.gdb** in ArcCatalog and notice the file has been added to the geodatabase. It should also open in ArcGIS.

5 Add the aerial photo to geodatabase

1. In ArcCatalog, navigate to the georeferenced Alamo aerial photo (Alamo.jpg) used in chapter 13. (Or you can access it via the CD provided with this book).

2. Right-click the file and select Export; then select Raster to Different Format.

3. The first field Input Features should already have the file you want to export to the geodatabase.

4. In the second field Output Location, use the browse button to navigate to **MyNewGDB.gdb** and then type a new name — let's call it **TheAlamo**. Click Save.

   Before moving to the next step, verify the path that it's saving to and make sure you are exporting to the geodatabase. The software has a tendency to connect to the wrong path here.

5. Click OK. It will take a minute to process.

6. Once it finishes, double-click **MyNewGDB.gdb** in ArcCatalog and notice that the file has been added to the geodatabase. It should also open in ArcGIS, but it will be so small it will be hard to see.

7. In the table of contents, right-click **TheAlamo** and select Zoom to Layer. The image and the street network should both be visible. If they are not visible together, you may be using the nongeoreferenced Alamo JPEG.

6 Add an Excel table to the geodatabase

1. In ArcCatalog, navigate to the social services file (**SocialServiceAddresses.xls**) used in chapter 8. (Or you can access it via the CD provided with this book).

2. Click the plus sign next to the file to expand the menu.

3. Right-click the worksheet **SocialServiceAddresses$** and select Export, and then select To Geodatabase (single).

4. The first field Input Features should already have the file to export to the geodatabase.

5. In the second field Output Location, click the browse folder icon to navigate to **MyNewGDB.gdb**. You will likely have to use the drop-down menu to select Folder Connections and then navigate to the folder where you created the geodatabase. Select the geodatabase (you do not give it a new name here) and click Add.

6. In the Output Table field, type a new name. Call it **Addresses**.

7. Click OK. It will take a minute to process.

8. Once it finishes, double-click **MyNewGDB.gdb** in ArcCatalog and notice that the file has been added to the geodatabase. It will also open in ArcGIS, but you won't be able to see it because it doesn't have any geographic attributes.

9. To see the data table, in the table of contents, click the List by Source icon and the data table will be visible in the table of contents. 🗐

Data Source Credits

Chapter 1
\ESRIPress\GIS20\Exercise_Data\Ch1\tl_2009_01_county.shp, courtesy of U.S. Census Bureau TIGER/Line

\ESRIPress\GIS20\Exercise_Data\Ch1\tl_2009_01_place.shp, courtesy of U.S. Census Bureau TIGER/Line

Chapter 2
\ESRIPress\GIS20\Exercise_Data\Ch2\tl_2009_01_county.shp, courtesy of U.S. Census Bureau TIGER/Line

Chapter 3
\ESRIPress\GIS20\Exercise_Data\Ch3\tl_2009_01_county-AlreadyProjected.shp, courtesy of U.S. Census Bureau TIGER/Line

\ESRIPress\GIS20\Exercise_Data\Ch3\tl_2009_01_county-Unprojected.shp, courtesy of U.S. Census Bureau TIGER/Line

Chapter 4
\ESRIPress\GIS20\Exercise_Data\Ch4\dt_dec_2000_sf3_u_data1.xls, created by the author

Chapter 5
\ESRIPress\GIS20\Exercise_Data\Ch5\tl_2009_01_county.shp, courtesy of U.S. Census Bureau TIGER/Line

Chapter 6
\ESRIPress\GIS20\Exercise_Data\Ch6\AgeJoined.shp, created by the author

Chapter 7
\ESRIPress\GIS20\Exercise_Data\Ch7\AgeJoined.shp, created by the author

Chapter 8
\ESRIPress\GIS20\Exercise_Data\Ch8\socialserviceaddresses.xls, courtesy of Dun & Bradstreet Inc.

\ESRIPress\GIS20\Exercise_Data\Ch8\tl_2009_48029_edges.shp, courtesy of U.S. Census Bureau TIGER/Line

Chapter 9
\ESRIPress\GIS20\Exercise_Data\Ch9\AgeJoined.shp, created by the author

\ESRIPress\GIS20\Exercise_Data\Ch9\Geocoding_Result.shp, created by the author

Chapter 10
\ESRIPress\GIS20\Exercise_Data\Ch10\tl_2009_48029_edges.shp, courtesy of U.S. Census Bureau TIGER/Line

\ESRIPress\GIS20\Exercise_Data\Ch10\XYData.xls, created by the author

Chapter 11
\ESRIPress\GIS20\Exercise_Data\Ch11\tl_2009_us_county.shp, courtesy of U.S. Census Bureau TIGER/Line

\ESRIPress\GIS20\Exercise_Data\Ch11\tl_2009_us_state.shp, courtesy of U.S. Census Bureau TIGER/Line

Chapter 12
\ESRIPress\GIS20\Exercise_Data\Ch12\tl_2009_01_county.shp, courtesy of U.S. Census Bureau TIGER/Line

\ESRIPress\GIS20\Exercise_Data\Ch12\tl_2009_01_place.shp, courtesy of U.S. Census Bureau TIGER/Line

Chapter 13
\ESRIPress\GIS20\Exercise_Data\Ch13\Alamo.jpg, courtesy of DigitalGlobe

\ESRIPress\GIS20\Exercise_Data\Ch13\tl_2009_48029_edges.shp, courtesy of U.S. Census Bureau TIGER/Line

Chapter 14

\ESRIPress\GIS20\Exercise_Data\Ch14\TargetArea.jpg, courtesy of Providence Community Housing

\ESRIPress\GIS20\Exercise_Data\Ch14\TargetArea.jpg.ovr, courtesy of Providence Community Housing

\ESRIPress\GIS20\Exercise_Data\Ch14\tl_2009_22071_edges.shp, courtesy of U.S. Census Bureau TIGER/Line

Chapter 15

\ESRIPress\GIS20\Exercise_Data\Ch15\AgeJoined.shp, created by the author

Chapter 16

\ESRIPress\GIS20\Exercise_Data\Ch16\tl_2009_01_county.shp, courtesy of U.S. Census Bureau TIGER/Line

\ESRIPress\GIS20\Exercise_Data\Ch16\tl_2009_01_place.shp, courtesy of U.S. Census Bureau TIGER/Line

Chapter 17

\ESRIPress\GIS20\Exercise_Data\ch17\AgeJoined.shp, created by the author

Chapter 18

\ESRIPress\GIS20\Exercise_Data\ch18\tl_2009_01_place.shp, courtesy of U.S. Census Bureau TIGER/Line

Chapter 19

\ESRIPress\GIS20\Exercise_Data\ch19\Alamo.jpg, courtesy of DigitalGlobe

Chapter 20

\ESRIPress\GIS20\Exercise_Data\ch20\Alamo.jpg, courtesy of DigitalGlobe

\ESRIPress\GIS20\Exercise_Data\ch20\socialserviceaddresses.xls, courtesy of Dun & Bradstreet Inc.

\ESRIPress\GIS20\Exercise_Data\ch20\tl_2009_48029_edges.shp, courtesy of U.S. Census Bureau TIGER/Line

Related titles from ESRI Press

A to Z GIS: An Illustrated Dictionary of Geographic Information Systems
ISBN: 978-1-58948-140-4

Written, developed, and reviewed by more than 150 subject-matter experts, _A to Z GIS_ is packed with more than 1,800 terms, nearly 400 full-color illustrations, and seven encyclopedia-style appendix articles about annotation and labels, features, geometry, layers in ArcGIS, map projections and coordinate systems, remote sensing, and topology.

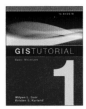

GIS Tutorial 1: Basic Workbook, Fourth Edition
ISBN: 978-1-58948-259-3

Updated for ArcGIS Desktop 10, _GIS Tutorial 1: Basic Workbook_ provides effective GIS training in an easy-to-follow, step-by-step format. By combining ArcGIS tutorials with self-study exercises intended to gradually build upon basic skills, _GIS Tutorial 1_ is fully adaptable to individual needs, as well as the classroom setting.

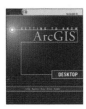

Getting to Know ArcGIS Desktop, Second Edition
ISBN: 978-1-58948-260-9

This workbook introduces principles of GIS as it teaches the mechanics of using ESRI's leading technology. Key concepts are combined with detailed illustrations and step-by-step exercises to acquaint readers with the building blocks of ArcGIS Desktop, including ArcMap, for displaying and querying maps; ArcCatalog, for organizing geographic data; and ModelBuilder, for diagramming and processing solutions to complex spatial analysis problems.

The ESRI Guide to GIS Analysis, Volume 1: Geographic Patterns and Relationships
ISBN: 978-1-87910-206-4

This book presents the necessary tools to conduct real analysis with GIS. Using examples from various industries, this book focuses on six of the most common geographic analysis tasks: mapping where things are, mapping the most and least, mapping density, finding what is inside, finding what is nearby, and mapping what has changed.

ESRI Press publishes books about the science, application, and technology of GIS. Ask for these titles at your local bookstore or order by calling 1-800-447-9778. You can also read book descriptions, read reviews, and shop online at www.esri.com/esripress. Outside the United States, contact your local ESRI distributor.